Eugen Sandow, Graeme Mercer Adam

Sandow on Physical Training

Eugen Sandow, Graeme Mercer Adam

Sandow on Physical Training

ISBN/EAN: 9783337365400

Printed in Europe, USA, Canada, Australia, Japan

Cover: Foto ©berggeist007 / pixelio.de

More available books at **www.hansebooks.com**

SANDOW

ON

PHYSICAL TRAINING;

A Study in the Perfect Type of the Human Form—*the Marvel of Anatomists, Sculptors, and Artists in the Nude; embracing the great Athlete's simple method of Physical Education for the Home, the Gymnasium, and the ... Training ...; preceded by a Biography dealing with the chief incidents in Mr. Sandow's Professional Career, his Phenomenal Prowess and Gladiatorial Skill, in Competitive Matches, Contests and Exhibitions; with Mr. Sandow's Scheme of Dumb-bell and Bar-bell Exercises, and his Views on the Physiology of Gymnastics, the Function of the Muscles, etc., etc.*

COMPILED AND EDITED, UNDER MR. SANDOW'S DIRECTION,

BY

G. MERCER ADAM,

EX-CAPT. QUEEN'S OWN RIFLES, C.M.

J. SELWIN TAIT & SONS
1894

TO

LIEUT.-COLONEL G. M. FOX,

HER MAJESTY'S INSPECTOR OF GYMNASIA

FOR THE BRITISH ARMY, ALDERSHOT,

I Dedicate this Work

IN GRATEFUL REMEMBRANCE OF MANY ACTS OF FRIENDLY

COURTESY, AND AS A TRIBUTE OF ADMIRATION FOR A

GALLANT SOLDIER AND A ZEALOUS ADVOCATE

OF PHYSICAL TRAINING ALIKE FOR THE

MILITARY MAN AND THE CIVILIAN.

EUGENE SANDOW.

NEW YORK, *January*, 1894.

SANDOW AT THE AGE OF TEN.

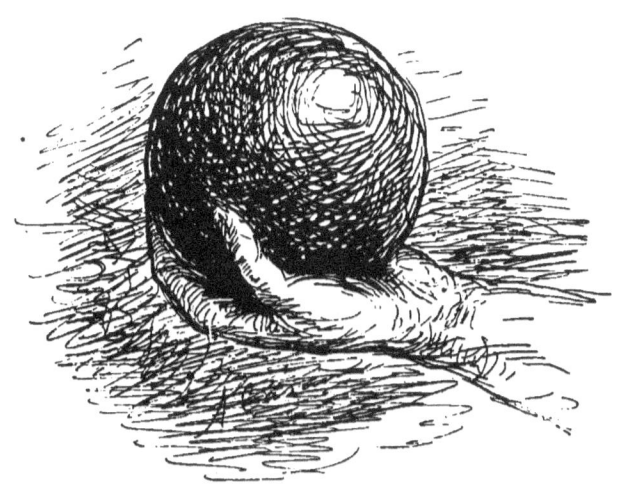

PREFACE.

THE following pages have been prepared under Mr. Sandow's direction and personal supervision. In the practical section appended to the narrative account of the great athlete's early amateur and later professional life, Mr. Sandow has furnished detailed instructions for the performance of his dumb-bell and bar-bell exercises and supplied the reader with a text-book which, he would fain hope, will be useful to the would-be athlete and to all who desire to attain perfect health, increased strength, and the full development of their physical frame.

Since the volume was put in type, further testimony, of a gratifying kind, to the value of Mr. Sandow's system of physical training has come to hand, in Captain Greatorex's courteous letter, to be found in the Appendix. It is regretted that the communication was not received in time to insert in the chapter to which it belongs—that on "Physical Culture in Relation to the Army." The letter forms a pleasant pendant, much prized by Mr. Sandow, to the one which appears in the chapter referred to, from Colonel Fox, H. M. Inspector of Military Gymnasia for the British army.

The illustrations to the practical as well as to the narrative portions of the book will, it is believed, add no little to its value. To the courtesy of Messrs. Sarony of New York, Morrison of Chicago, and H. Roland White of Birmingham, England, the publishers are indebted for permission to reproduce the photographs.

The Editor takes advantage of this prefatory note to acknowledge his obligations to Mr. Sandow and his pupil, Mr. Martinus Sieveking ; to Mr. W. T. Lawson, member of the New York Athletic Club ; to Dr. D. A. Sargent of the Hemenway Gymnasium, Harvard University ; to Dr. Everett M. Culver of New York ; to Dr. W. Theophilus Stuart of Toronto, Canada, and to the Publishers, for courtesies received during the preparation of the work.

NEW YORK, February 1, 1894.

CONTENTS.

INTRODUCTORY.

CHAPTER I.

A PLEA FOR PHYSICAL EDUCATION.

                                                         PAGE

Consummate beauty of physical form—Knowledge possessed by the ancients in relation to physical training—The jar and fret of modern business life—Health rather than strength the great requisite of the times—Sports and pastimes of the people—Appurtenances of our gymnasia too costly and elaborate—All exercises should be performed on the ground—Attention to chest development—The prolific causes of disease and physical degeneracy 1

CHAPTER II.

SANDOW, A TITAN IN MUSCLE AND THEWS.

Sandow a study for the physiologist and anatomist—For four years the lion of London—Crowned heads pay him honour—Notable scientists give testimony as to his muscular power and physical endowments—His system of physical training adopted for the British army—Examined by Dr. Sargent, of Harvard—Mighty deeds of ancient story—Emulating effects of these heroic acts—Sandow comes to know his own power 12

CHAPTER III.

SANDOW'S BOYHOOD AND EARLY LIFE.

At birth nothing of a prodigy—Inherits simply a healthy and normally well-developed frame—His student days—Attached to the gymnasium and the

CONTENTS.

circus—Becomes notable as a wrestler—Visits Rome with his father and admires classical sculpture—Decline of the physical ideal—Quarrels with his father and runs away from home—Enters University of Gottingen—Studies anatomy at Brussels—Meets Atilla—First public exhibitions...... 21

CHAPTER IV.

SANDOW AS A STRONGMAN IN HOLLAND.

Sandow dependent upon his own resources—Arrives at Amsterdam and seeks employment as a strongman—Daring scheme to advertise himself—Hard up, and takes a cabman into confidence—Wrecks the machines for testing strength throughout the city—A thousand guilders reward—Arrested; amusing scenes at the police station—Released, and makes the fortune of a hotel-keeper—Receives his first engagement at a theatre—First visit to London—Accident to Atilla, and is thrown out of employment—Goes to Paris—Fruitless efforts to get an engagement—Startles a professor at the Academy of Arts with an exhibition of his strength—Earns 200 francs as a model—Meets François and joins him in pantomime..................... 28

CHAPTER V.

SANDOW AS A WRESTLER IN ITALY.

Visits Rome and gives exhibitions in the Colosseum as a wrestler—Performs mighty feats of strength—Wrestles with Bartoletti and wins 1,000 francs—Achieves fame and has King Humbert and his court as admirers—Gift from the king—Visits Emperor Frederick by command at San Remo—Astonishes the Kaiser by an exhibition of his powers—Receives a ring from Frederick—Pathetic words of the dying Emperor at the leave-taking. Wrestling matches at Florence, Milan, and Naples—Contest with three trained athletes and puts all successively on their backs—Wins 5,000 francs—Buys a home at Venice—Hurts his arm in a wrestling contest—Retorts with a loving embrace—Attracts the attention of an English painter—Makes him the subject of a study—Tells him of Samson's challenge—Starts post-haste for the British metropolis............................. 35

CHAPTER VI.

SANDOW WINS HIS FIRST LAURELS IN LONDON.

Sandow takes London by storm—Pen portrait of the young athlete—Lifts Samson's gage of battle and beats his pupil Cyclops—Wins the £100 wager

CONTENTS.

—Great feats of strength at the Royal Aquarium—The London *Sportsman* on the contest—Accepts Samson's £500 challenge—London disillusionized. 43

CHAPTER VII.

DEFEATS SAMSON AT THE WESTMINSTER AQUARIUM.

Strong men in rivalry—Uproarious night at the Aquarium—Sandow flies the blue-peter of success—Exciting scenes at the contest—Samson theatrical and querulous—Great talkee-talkee—The *Daily News* on the affair—Sandow declared winner of the £500—Relative merits of the two athletes' feats of strength—Fillip given by the contest to athletics—Engagement at The Alhambra—Royalty honours Sandow.................................. 54

CHAPTER VIII.

SANDOW IN SCOTLAND AND AT THE CENTRES OF INDUSTRIAL ENGLAND.

The Press and "the War of the Titans"—Sandow at The Alhambra—Tour of the Provinces—Sandow in Scotland—Repertoire of feats—Exhibition of mountains of muscle... 65

CHAPTER IX.

WITH GOLIATH AT THE ROYAL MUSIC HALL, HOLBORN.

The two giants, Sandow and the Aix-la-Chapelle quarryman—Crowded audiences—Varied programme of entertainment—Lifting 500 lbs. with one finger—At the London Pavilion with Loris—Phenomenal feats of strength 71

CHAPTER X.

ANOTHER STRONGMAN CONTEST.

The *Morning Post* on the match with "Hercules" McCann—Inexplicable issues of the contest—The Press on the miscarriage of justice—Wins £50 wager for lifting a 250-lb. weight from the shoulder................. 77

CHAPTER XI.

SANDOW BREAKS ALL RECORDS.

Wins the gold championship belt of the London Athletic Institute—Great right and left hand work—Breaking Hercules's record—Making three great records—At Birmingham and Liverpool.................................. 83

xii   CONTENTS.

CHAPTER XII.

PHYSICAL CULTURE IN ITS RELATION TO THE ARMY.

PAGE

Military circles interested in Sandow—Training depots take up his system of light dumb-bell exercise—Surgeon-Major Deane's Lecture at Woolwich—Sandow 'an object lesson in Gymnastic Anatomy'—Report of the London *Lancet*—Colonel Fox, H. M. Inspector of Military Gymnasia, endorses Sandow's methods of Physical Training...................... 89

CHAPTER XIII.

SANDOW "AT HOME" AND ABROAD.

A case of " bringing down the house "—Sandow *chez lui*—Risks of housing a strongman lodger—A holiday in Paris—An unpleasant rencontre—A pugnacious Frenchman—Severe chastisement of the aggressor—Sequel in London—Presented with a valuable chronometer—Tracking a brace of thieves at Nice—Sandow his own law-enforcer........................ 98

CHAPTER XIV.

SANDOW IN THE NEW WORLD.

Accepts American engagements—Opens at the Casino, New York—The New York *World* on Sandow—Sandow's great hitting power—His increasing strength—Interviewed by the New York *Herald*—Holding up three horses 105

CHAPTER XV.

SANDOW AS A PHYSIOLOGICAL STUDY.

Sandow as a physiological study—Examined by Dr. Sargent, of Harvard—The "strongest man measured "—Wonderful abdominal muscles—Ingenious electrical tests—Speed in delivering a blow..................... 119

CHAPTER XVI.

SANDOW SPEAKS FOR HIMSELF.

HIS VIEWS ON PHYSICAL TRAINING, DIETING, BATHING, EXERCISING, ETC.

Physical perfection of the great athlete—The culmination of a system which will enable the weakest to become strong—Predisposing causes of San-

CONTENTS. xiii

dow's physical strength—A reporter's interview—How Sandow became muscular—His effective system—Further chat with the strongman—Results of his training—His faith pinned to the use of light-weight dumb-bells.. 129

CHAPTER XVII.

THE PHYSIOLOGY OF GYMNASTICS.

Mr. Sandow's introduction to his practical exercises—His views on the theoretic and practical bearing of physical training—Influence of bodily exercise on the human organism—A symmetrical and all-round development—Exercise in fresh air—Dumb-bell and bar-bell exercises recommended—Ineffective and vicious systems of training—Correct habits of breathing. 140

CHAPTER XVIII.

HYGIENIC AND MEDICAL GYMNASTICS.

The rationale of gymnastics—Effect of exercise in beautifying women—Prejudice, indifference, and delusion—The bugbear of training—Hygenic effects of exercise—Muscular exercise as an aid to digestion—Dieting and food—The coarser meats the best for sustenance—How Sandow passes the day—Influence of exercise on the mind—Perils of over-exercise...... 152

CHAPTER XIX.

EXERCISE AND THE BODILY FUNCTIONS.

Neglect of Exercise as an agent and promoter of health—The ambition commendable to be healthy and strong—The inter-relation of body and brain—Mr. Sandow remarkable as a human motor—The secret of heavy-weight lifting—The problem of obesity solved—The skin and its functions....... 170

CHAPTER XX.

THE CHIEF MUSCLES, WHERE THEY ARE SITUATED, AND WHAT THEY DO.

The muscles actively concerned in the movements of the body—The voluntary and involuntary muscles—Those that are chiefly affected by muscular exercise—The muscles of the upper chest, back, shoulders and arms—The chief muscles of the lower extremity—the hip, thigh, and leg............ 178

CONTENTS.

PRACTICAL EXERCISES.

PAGE

Prefatory:—Instructions to young would-be athletes—Hints to pupils and instructors—Preliminary free movements for rendering the muscles and joints supple... 199
Light-weight dumb-bell exercises.. 208
Heavy-weight dumb-bell exercises... 218
Bar-bell exercises 227
Finger lift, stone lift, and harness-and-chain lift........................ 232

SANDOW'S PHYSICAL TRAINING LEG-MACHINE.

Description of, and suggested methods of using it...................... 235

   Jendix A.—Table of Food substances and their nutritive value........ 239
    B.—Anthropometric chart of Mr. Sandow's measurements.... 241
    C.—Table showing results of muscular development of a pupil of Mr. Sandow's, after three months' practice of his systematized exercises (see photo. of pupil) 242
    D.—Letter from Assistant-Inspector of Military Gymnasia for the British army.... 243

LIST OF ILLUSTRATIONS.

ENGRAVINGS ILLUSTRATING THE POSINGS, ETC.

	PAGE
Portrait of Mr. Sandow, in Street Attire, with Autograph Sig.	*Frontispiece*.
Sandow at 10 years of age	vi
Sandow in a series of 4 Club Studies	28
Forearm Studies: Sandow's Flexed Arm, showing Deltoid and Serratus Magnus Muscles (two illustrations)	29
Forearm Studies: Sandow's Flexed Arm, showing Biceps and Triceps Muscles (two illustrations)	89
Sandow seated, showing abdominal muscles	98
Sandow (full figure, lateral position), Arm Flexed	112
Sandow in a series of 4 Classical Poses	113

ENGRAVINGS ILLUSTRATING THE MUSCLES.

Athlete in the Pose of elevating the Ring-and-Ball 152
Skeleton of Athlete (full figure) 170
Muscles of Athlete (anterior aspect) 153
Muscles of Athlete (posterior aspect) 171
Muscles of the Flexed Arm (anterior, posterior, and lateral aspects) 185
Muscles of the Trunk, Shoulder, Extended Arms and Flexed Leg 186
Muscles of the Extended Leg (anterior, posterior, and lateral aspects) 196
Portrait of a pupil of Mr. Sandow (Mr. Martinus Sieveking) 243
Dr. Sargent's Anthropometric Chart of Sandow 241

ENGRAVINGS ILLUSTRATING THE EXERCISES.

Light-weight Dumb-bells.

Nos. 1–4. For developing the arm flexor and extensor muscles 210
5 *a.* Chest-opening exercise (first position) 212

ILLUSTRATIONS.

	PAGE
Nos. 5 b. Chest-opening exercise (second position)	
6. For developing the trapezius and latissimus dorsi muscles	213
7. For increasing the mobility of the shoulder-joints	
8 and 9. For making flexible the muscles of the wrist and forearm	
11 a and b, 12. Lunging exercises, for developing the shoulder and arm muscles and those of the chest and sides	214
13 a and b. Chest-expanding exercise	216
14 a and b. Chest expanding exercise, with machine resistance	217
15 a, b, and c. For strengthening the muscles of the abdomen and preventing obesity...	218

Heavy-weight Dumb-bells.

Nos. 18, 19. How to lift by one hand from the ground to the shoulder.....	220
20, 21, 22, 23, 24. Illustrating one-handed slow-press from the shoulder...	222
25, 27. One-hand swing-lift from the ground overhead	224
28. Slow-lift from the ground to the shoulder...................	
29. Snatch ring-and-ball lift from the ground overhead	224
30, 31. Two-handed lift from the ground to the shoulder	
33, 34. Holding-out exercise at arm's length with both hands	226

Bar-bell Exercises.

Nos. 35, 36. Illustrating one-handed lift from the ground to the shoulder	227
37. One-handed snatch-lift from the ground overhead	
38 a, b. Bar-bell exercise for one hand	229
38 c and d. Bar-bell exercise for two hands	230
39 and 39 a. Slow bar-bell lift for developing the muscles of the forearm and wrist..	231
40 a. One-handed bar-bell lift, upright position	232
Two-handed bar-bell lift to the shoulder, upright position ...	

Miscellaneous Exercises.

Nos. 43. Illustrating stone-lift from the ground for one and two hands	233
44. Harness-and-chain lift from the ground..........................	234
45 a, b, c. Illustrating leg-machine exercises........................	237
45 d and e. " " "	236

SANDOW ON PHYSICAL TRAINING.

I.

A PLEA FOR PHYSICAL EDUCATION.

IN spite of the increasing value of individual life —the distinctive mark of the civilization of our time—little has as yet been done, on large lines at least, to secure for the masses of the people who do the work of the world that degree and maintenance of physical well-being implied in the phrase, " a sound mind in a sound body." For those even whom we are pleased to call " the flower of our population," we have systematically and intelligently done next to nothing in the way of physical culture. Only in recent years has physiology been put on the curriculum of our public schools and the young have been enabled to get some inkling into the frame-work of their bodies and the physical conditions on which organic life is held. Whether this knowledge, in the main, goes beyond an appreciation of the necessity for air, light, food, clothing, and cleanliness, as conditions essential to health, may be greatly doubted. What is remembered of the theoretic laws of health when schooldays are over, is, if we except the case of the comparatively small contingent that goes on to the study of medicine as a profession, of little value in the practical government of our bodies. Even what we have picked up about sanitation is generally lost before we have well entered upon manhood, or is effectively and grimly set at naught in our homes by the plumber. Where physiology has been properly taught, we may not all be as heathen in our knowledge of the requisites of health. In a few fortunate instances, the youth may know something of the processes of waste and renovation in the body; but how those processes work to the best advantage and show their most beneficent results under the systematic exercise of the muscular system, is, admittedly, given to but few of us fully to appreciate or wisely to understand. Even the ancient Greeks, noted as they were for their fine physical development, grace and symmetry of form, groped largely in the dark regarding many things which modern physiological science has now made plain. This is well understood; but, with the higher knowledge that modern science has brought us, how indifferent has been our approach to

THE CONSUMMATE BEAUTY OF PHYSICAL FORM

for which the Greek—especially the Athenian athlete—was famed. Greek and Roman alike knew, in a high degree, the value of bodily exercise, and in their competitive games, as well as in their training for war, adopted a system of physical education which produced wonderful results. They knew nothing, however, of biology and the marvel of the body's cell-structure, the key which, it may be said, has opened to a modern age the doors of its microscopic vision and revealed almost the secret of life itself, with its ever-recurrent motions of waste and renewal. They did not know, as Mr. Archibald Maclaren, the great English authority on Physical Education, has observed, "that man's material frame is composed of innumerable atoms, and that each separate and individual atom has its birth, life, and death; and that the strength of the body as a whole, and of each part individually, is in relation to the youth or newness of its atoms. Nor did they know that this strength is consequently attained by, and is retained in relation to, the frequency with which these atoms are changed, by shortening their life, by hastening their removal and their replacement by others; and that whenever this is done by natural activity, or by suitable employment, there is ever an advance in size and power, until the ultimate attainable point of development is reached. They simply observed that the increased bulk, strength, and energy of the organ or limb is in relation to the amount of its employment, and they gave it employment accordingly."

This, in the main, was the sum of knowledge possessed by the ancients in relation to physical training; yet unscientific —as we now understand the term—as it was, its results were wonderful in promoting strength and activity. Of course, in giving themselves so ardently to physical education, the Greeks and Romans must have observed much else, as the results of muscular exercise, that was beneficial to the youth in training. Though they had little knowledge of the why and wherefor in physiological law, they saw its gratifying effects and so betook themselves, with increasing national enthusiasm, to the exercises of the gymnasium and the campus. The physiological action on the lungs and the blood produced by quickened respiration, incident to regular periods of muscular exercise, they might not know; but they saw clearly its health-giving results, on the mind as well as on the body, though no doubt, with them as with us, it was the few only who were qualifying themselves for the service of war who had the benefit of this experience in training. Interest in the physical well-being of any beyond those who were designed to bear arms, there was none in either Athens or Rome. Outside of that favoured class there was no public provision for physical education; though there were always patriotic and high-spirited youth whom the thirst for distinction drew into the competitive arena to take part in wrestling contests, swimming matches, chariot racing, and other national sports and games. With us, of recent years at least, physical training has gone beyond the parade-ground or barrack-room of the soldier. It has happily found its way into our schools and colleges, and, in a few of them, at any rate, it takes a place on the curriculum hardly inferior to that assigned to intellectual studies. Of late years, also, provision has specially been made for it by athletic clubs and other organizations for recreation, of a private or corporate character, with results that have gone far to neutralize the physical deterioration that in our over-competitive age is incident to

THE JAR AND FRET OF BUSINESS LIFE.

Theoretically, at least, we all pay tribute to the value and importance of physical education. We admire physical strength and beauty, and recognize, though only faintly as yet, the inter-relation of mind and matter. We know, moreover, that a healthy, active brain is sadly handicapped by an ill-developed, sickly body. We see around us every day of our lives masses of our race of imperfect growth and unsound constitution, and almost daily the lesson comes home to us of the break-down of some friend or acquaintance, whose weakness of body could not withstand the mental and bodily strain in the struggle of life. Yet it is not strength, so much as health, that is the crying want of the time. It is stamina, and the power, in each of us, to do our daily work with the least friction and the greatest amount of comfort and ease. Only the few are called upon, like the great traveller or the soldier in a campaign, to endure protracted fatigue and encounter serious obstacles in nature or severities of climate, from which most of us shrink, and for the undertaking of which few of us have either the will-power or the courage. "A small portion only of our youth are in uniform," observes the authority we have already quoted; "but other occupations, other demands upon mind and body, advance claims as urgent as ever were pressed upon the soldier in ancient or modern times. From the nursery to the school, from the school to the college, or to the world beyond, the brain and nerve strain goes on—continuous, augmenting, intensifying. Scholarships, competitive examinations, speculations, promotions, excitements, stimulations, long hours of work, late hours of rest, jaded frames, weary brains, jarring nerves—all intensified and intensifying—seek in modern times for the antidote to be found alone in physical action. These are the exigencies of the campaign of life for the great bulk of our youth, to be encountered in the schoolroom, in the study, in the court of law, in the hospital, and in the day and night visitations to court and alley and lane; and the hardships encountered in these fields of warfare hit as hard and as suddenly, sap as insidiously, destroy as mercilessly, as the night-march, the scanty ration, the toil, the struggle, or the weapon of a warlike enemy.

"Yes, it is *health* rather than *strength* that is the great requirement of modern men at modern occupations; it is not the power to travel great distances, carry great burdens, lift great weights, or overcome great material obstructions; it

is simply that condition of body, and that amount of vital
capacity, which shall enable each man in his place to pursue
his calling, and work on in his working life, with the greatest
amount of comfort to himself and usefulness to his fellow-
men. How many men, earnest, eager, uncomplaining, are
pursuing their avocations with the imminency of a certain
breakdown ever before them—or with pain and weariness,
languor and depression, when fair health and full power
might have been secured, and the labour that is of love, now
performed incompletely and in pain, might have been performed with completeness and in comfort."

Nor is the remedy hard to apply or likely to be at all
doubtful in its results. It is Nature's own panacea—the
remedy, as we have seen, which the nations of antiquity,
intelligent and highly civilized as they were, found effective
in war as well as conducive to the health and vigour of youth.
But physical strength was not only "the veritable God of
antiquity;" it was also the pride and idol of the Middle Ages.
At the latter era, the tilting-field and tourney-ground took
the place of the Campus Martius and the gymnasium. There
the chivalry of the time disported itself in jousts and feats
of horsemanship, while the village-green gave encouragement
to wrestling matches and the varied sports which are noted
among England's manly national games. We in the New
World are inheritors of many of these playful incitements to
bodily vigour, to which we have added others, characteristic
of our climate and people, but all helpful in their way in the
up-building of a lusty frame. Valuable, however, as are these

SPORTS AND PASTIMES OF THE PEOPLE,

they are only recreative exercises and, for the most part, fitfully
indulged in. Moreover, they are confined, as a rule, to the
school-age, and are too often dropped when the youth passes

into the first stage of manhood. It is well known, also, that they develop only the lower limbs, or the lower limbs and the right arm, leaving without its meed of exercise the left arm and upper portions of the trunk. This incomplete and imperfect unfolding of the human body it should be the design of intelligent methods of physical training to correct and to supply with the needed exercises, so as to bring about a uniform and harmonious development. Lacking this, there is seen faulty growth and weak or distorted conformation in an otherwise healthy and well-constructed frame.

In the following pages, the narrative of the career of an enthusiast in athletic pursuits, it is the design of Mr. Sandow, as well as the modest purpose of the writer, to show how effective can be even simple methods of muscular training, when scientifically imparted, in raising the human body to a high plane of physical perfection, and in making it better fitted for the all-round, every-day work of both the manual and the intellectual toiler. In physical education, as in every other laudable ambition, there are few royal roads to the signal and satisfactory attainment of one's ends. Here the sciolist, or the ill-equipped instructor, can of course make a show of juggling, and hump the muscles in indiscriminate ridges, without much reference to their practical uses, and with little benefit to the health, vigour or permanent well-being of the deluded pupil whom he affects to train. This, of course, is folly. In all our aims after physical education the great thing to bear in mind is to avoid ambitious and elaborate efforts at bodily training. The ancient Greeks and Romans would have laughed at our extensive array of apparatus,—the appurtenances of our modern gymnasia—on which we foolishly lavish large sums of money, often only to be looked at, or used for harm rather than for good. Another point is this : see that your training be not only simple but effective. In its scope let it be thorough. Physical education, as we have already hinted, is too often and incompletely directed to the accomplishment of one or two feats—notably those wrought by the exterior muscles by the use of the apparatus ordinarily in vogue in our gymnasia—without reference to the vast net-work of interior muscles, which have so much to do with bearing the strain of arduous gymnastic exercise, and have their important, set functions in the vital seat of the system. As these interior muscles are brought into harmonious play with the connected exterior folds of tissue, the athlete may pursue his exercises safely; if they are not so brought into play, as too often happens, then a break-down may be expected, and dire, often, is the result. To obviate this, Mr. Sandow's stringent caution cannot be too strongly impressed, on the young gymnast particularly, viz., that

ALL EXERCISES SHOULD BE PERFORMED ON THE GROUND,

where nature intended the human animal to find his habitat, and there to stand erect. He also wisely enjoins the use of dumb-bells of only 5 lbs. in weight, for the earnest and systematic manipulation of these, he affirms, is sufficient for the due development of all the muscles and groups of muscles appertaining, at least, to the upper part of the body; while by confining the would-be athlete to these medium-sized bells no risk of injury is run, and the average man can be kept in the perfection of health. This result will be the more assured, if the pupil-in-training will make himself intelligently acquainted with the anatomical arrangement and disposition of his muscles, and acquire some practical knowledge of physiological science. For the development of the lower limbs, Mr. Sandow has constructed and patented a simple apparatus which, he claims, is, with the light-weight dumb-bell, all that the athletic devotee needs for the vigorous up-building of his body. The mechanical contrivance referred to will be found admirable for exercising the adductor muscles of the leg. Its usefulness need hardly be pointed out, to those, at any rate, who have seen Mr. Sandow in what is familiarly called the Roman Column feat, and have observed what muscular strength he possesses in his lower limbs (though in the performance of this feat other muscles than those of the lower limbs are called more into play), which are kept in training partly by the use of this ingenious invention.

Of course, the mass of humanity, even of those who do the heaviest part of the world's work, are not likely, whatever time they can give to physical culture, to become Titans in strength. Nature is wont to be churlish when she is expected to make prodigies of us all in either physical or intellectual vigour. Yet nature is no niggard in placing at the disposal of the race, at least, the raw material out of which it may fashion both vigorous minds and healthy bodies. The trouble is that our modern methods of education, for the most part, do not lead to mutual and concerted action in the training of these dual parts of our being. The mistake is the more serious when we realize how great is the influence on the mind of a physically well-developed body. Equally important is the realization of the truth, that a strong man, well-trained, can put his strength to an incalculably greater advantage than a man of like vigour whose physical powers have not been cultivated. Even a superficial perusal of the following pages can hardly fail to attest, and, it may be, impress this lesson.

But the prime lesson for all, is to seek to raise the individual physical strength, which, unquestionably, is much lower for the race than it ought to be. By raising the physical standard in the unit, time and training will accomplish like results for the race. Nor are we without encouragement in seeking, in either unit or race, an improvement in physique; for Mr. Sandow, who is what he has made himself by following his own simple system of muscular training, is a striking illustration of the power of expansion latent in the human frame, and which in the most of us is capable of development. Physically, Mr. Sandow is, of course, of more than normal girth, as well as of exceptional strength of chest, loin and limb ; but under favouring conditions of exercise and training many might attain to the same measure of physical development, while none need despair of making some gratifying approach to it. We repeat, however, that health, rather than muscular strength, should be the chief object of physical training. To most of us, engrossed in the ordinary avocations of life, and necessarily confined by the conditions of our occupations to sedentary habits, the main consideration must be the degree in which we can best perform our work, with the utmost attainable freedom from friction or bodily ailment. In Mr. Sandow's scheme of training he properly gives much

ATTENTION TO CHEST DEVELOPMENT,

since, unless the heart and lungs have room for their natural and active play, it will matter little either how large or how strong may be the legs or arms. A narrow or weak chest is not only in itself a serious bodily defect, but it invariably conduces to an inferior physique. This has been well illustrated by facts recently gathered by Dr. G. W. Hambleton, President of the Polytechnic Physical Development Society, of London, who has made many years' researches into the vocations which induce weak lungs and contracted chests. To the neglect of a proper chest development, says this authority, is due the large reduction from the numerical strength of the British army, a reduction which is not only a national weakness, but the occasion of much financial loss, in the annual invaliding and death of so many otherwise effective men from the ranks. Benefit societies and life assurance companies, Dr. Hambleton also computes, lose an enormous sum yearly from the same inciting cause, which might be largely removed, were the tendency of the habits and the surroundings of the insured such as to secure increased breathing capacity. Indifferent breathing power, and the lack of fresh air and proper muscular exercise, are but too certainly the prolific causes of disease and physical degeneracy. Well will it be when the masses recognize and act upon this palpable truth. Well also will it be when our instructors make an effort to raise the prevailing type of chest to a more efficient standard of excellence.

What is further to be said on this important subject, and especially on the topic of vital interest to the youth-in-training —the practical bearing of muscular exercise on the health and strength—will be treated of in a later chapter in the technical division of the work, with the benefit of Mr. Sandow's own experience as a self-trained athlete and preceptor in the science of physical culture.

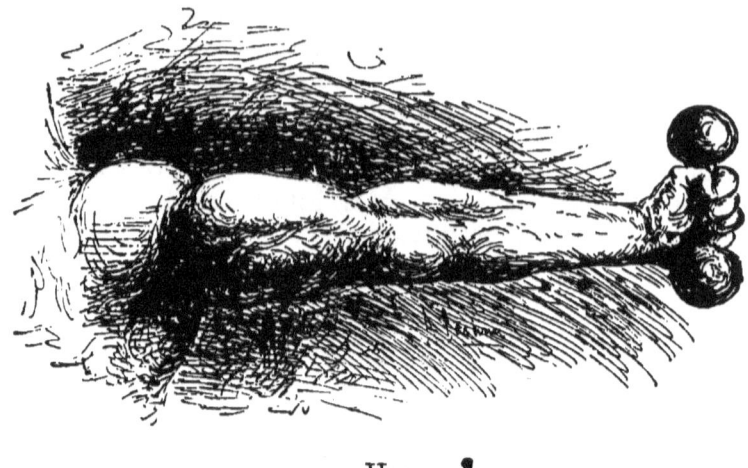

II.

SANDOW, A TITAN IN MUSCLE AND THEWS.

SANDOW, in the ideal perfection of his physical manhood, as he now appears, is a highly interesting and inspiring study for the physiologist and the worshipper of Titanically-developed muscle and thews. His athletic prowess ranks him with the heroes who are credited with doing mighty deeds in the Homeric age. Our modern times have produced no one, it is not too much to say, more perfectly equipped than is this young Prussian, either as an all-round athlete, or as an example of what musclar training can do in developing to perfection the human form and achieving the classical ideal of physical beauty. When, but a few weeks ago, he came to the New World, it might have been supposed—and the hyperbole in the present case is pardonable—that the advance-guard of a new order of physical beings had descended on our planet. Not only the ubiquitous reporter, but native strong men, and even experienced and widely-read physiologists, waxed eloquent in descanting on his points. But Eugene Sandow, on his advent in New York, neither fell romantically from the clouds nor came among us without record of his past doings or passport to public appreciation and favour. Young as he still is, he had been for four years the lion of London, the sensation of the time in the English Provinces, and was known to have been the hero of a hundred wrestling and gladiatorial contests on the Continent of Europe. In these matches he had beaten all competitors and won the hoarsely-shouted acclaim, with the more substantial awards of favour, of the sport-loving populace in the chief pleasure cities of the Old World.

CROWNED HEADS HAD PAID HIM HONOUR,

even royalty and the aristocratic youth at courts had been his pupils; while his name was everywhere a household one among all classes of the people. Anatomists of world-wide fame lovingly dwelt on his wonderfully developed frame before delighted students in the dissecting room, and sculptors and artists eagerly bid against each other to secure him as a model.

Nor are we without accredited testimony, from notable savants, as to the physical endowments of the great athlete. Professors Virchow, of Berlin, Rosenheim, of Leyden, and Vanetti, of Florence, have expressed this opinion, that Sandow, from an anatomical point of view, is one of the most perfectly-built men in existence. This judgment has been authoritatively endorsed by scores of English medical men, of high repute in their profession, as well as by hundreds of professors and well-known experts in the science of physical education. Army surgeons and chiefs in the training schools, in the great English depots at Woolwich and Aldershot, have also given unqualified testimony to Mr. Sandow's prowess and to the unprecedented results of his methods of training. In December of last year (1892), at the gymnasium of the Royal Military Academy, Woolwich, Surgeon-Major Deane, of the Medical staff, made Sandow the interesting theme of a lecture, notable, not only for its inherent merit, but also from the fact that the great athlete was present and afforded in his person, to the astonished cadets, a practical object-lesson in gymnastic anatomy.

HIS PHYSICAL TRAINING SYSTEM ADOPTED IN THE BRITISH ARMY.

In military circles throughout England, Mr. Sandow has been paid similar compliments, and has had the honour of having his system of physical training recommended for use in the training schools of the British army, through the agency of Colonel Fox, Inspector of Gymnasia at Aldershot, an enthusiastic admirer of Sandow, and a warm friend.

Since his arrival on our shores, Sandow has been the recipient also of not a little interested scientific attention, and been the subject of much wonder and admiring comment. In his exhibitions at the New York Casino, in the Tremont Theatre, Boston, and at the Trocadero, Chicago, he has drawn, nightly, thousands, the sincerity and heartiness of whose plaudits have emphasized the wonder and dexterity of his feats. Nor have athletes, amateur and professional alike, been either backward or grudging in their praise; while to anatomists and the medical faculty in general, Sandow appears—if one may venture the phrase—as a standing miracle. The New York Athletic Club have also paid him the undisguised tribute of admiration, one of its distinguished members having spoken of him as "the most perfectly-developed man he had ever seen." Another member of the Club remarks : "I have seen athletes with almost as big muscles, but never one with the all-round development Sandow possesses. There is nothing abnormal, moreover, in his development. The nearest approach to a deformity, if a natural muscular development may be termed a deformity, is in the abdominal muscles. The like of these I have never before seen in a human being."

Dr. Dudley A. Sargent, Director of Applied Anatomy, Physical Training, and Personal Hygiene, at the Hemenway Gymnasium, Harvard University, has compiled an interesting Anthropometric Chart of Sandow, recording accurate measurements of the different parts of the athlete's anatomy, and prepared a professional paper on him for the Press. In the latter he observes :

"SANDOW IS THE MOST WONDERFUL SPECIMEN OF MAN I HAVE EVER SEEN.

He is strong, active and graceful, combining in his person the characteristics of Apollo, Hercules, and the ideal athlete."

In recording these tributes to Sandow's amazing physical powers and phenomenal development of muscle, it is not the purpose of this volume, however, to set before the reader a mere panegyric, or to treat Sandow as a prodigy for exhibition purposes only. Far otherwise, as we hope shortly to show, is the design of this work.

THE MIGHTY DEEDS OF ANCIENT STORY.

In all ages there have been some few men possessed of unique physical power and great muscular development. Maximinius, the murderer of the Emperor Alexander Severus, is said to have been able to draw fully-laden carts and waggons without much effort, to crumble stones between his fingers, and tear young trees apart with his strong hands. Cæsar Borgia is also credited with the possession of great strength, for it is affirmed of him that he could knock down a bullock with one blow of his fist. A certain centurion in the Emperor Augustus's body-guard, tradition has it, could, by the sole strength of his arms, bear up a waggon laden with two hogsheads of wine until all the wine was drawn out. It is moreover said of him that he could carry a mule on his back as easily as he could carry a child, and stop a chariot when the horses were in full gallop. But we need not go back for such manifestations of great strength to a mythological age, or seek for them only among the heroes of antiquity. We have all read, or heard, of the Venetian athlete who, though small of stature, could break the thickest shank-bone of oxen upon his knees; of the German, Le Feur, in the sixteenth century, who could carry a pipe of wine on his shoulders; of John Bray, the Cornishman, who could carry the carcass of an ox a furlong; of Libeski, a Polander, who at Constantinople, in 1581, lifted a piece of wood which twelve men had enough to do to raise from the ground and bore on his breast a mighty stone which ten men had, with much effort, rolled thither. Nor do we forget the Scottish Highlander who, not long ago, used to uproot young oaks from the earth, cast Highland steers, and harnessing himself with horse-breeching raise a ton weight; or Topham, the strong man of the last century, who, with the aid of leathern straps passed over his shoulders, with chains attached, could lift three hogsheads of water, weighing 1836 lbs., and support on his body four men, each weighing fourteen stone.

THE EMULATING EFFECT OF MIGHTY DEEDS.

The ancient classics give us well-nigh a surfeit of mighty heroes, whose deeds have been sung in noble epic or recited in stirring story. If many of these deeds are mythical, the classical student has not the less enjoyed the literary qualities in the story and the story-teller ; nor have daring spirits, in the ages since, failed to find in both a stimulus to the accomplishment of feats of like prowess. In the swing and spirit of their telling, ardent natures have often caught fire, and done many a noble deed by emulating the spirit which even the recital of noble deeds inspire. England's battles have been won, it is a familiar saying, by the muscle-training which her youth acquire on the playgrounds of her great Public Schools. What they drink in, as with their mother's milk, of deeds of renown in their own noble history, as well as in the tradition of that of other nations—Greek, Roman, Teuton, and Scandinavian—may well fire the young heart to deeds of high emprise and great valour.

No lustre is so great, we know, as that which gilds the doing of a great deed. Back, however, of the doing of it, must be the courage which a consciousness of the ability to do the deed inspires. In this lies the moral value of physical training. We do not say, of course, that the intrepid mind waits to reason before throwing itself into the breach in the moment of jeopardy or peril. But is it not folly to hazard life in the performance of an act for the doing of which one has not the physical power, though one may have the courage? The man who is himself no swimmer will but court his own fate should be seek to save another from drowning. He who is most likely to stop a runaway horse in a crowded thoroughfare is the man who has both the muscle and the pluck to risk life in the effort. Nor is it safe to say that emergencies are infrequent for the instant

EXERCISE OF A STRONG MAN'S POWER AND WILL.

The student of martial history, at least, will not need to be reminded of this. Turn the dial of time back a few hundred years, and he will recall how often the fortunes of battle depended upon the deft prowess of a single arm. Nor is the fact less true of our own time. One can hardly go into the thronged streets of our cities, or board a crowded steamboat, on pleasure bent, without being confronted with an emergency which may call our whole strength and courage into instant action. Mr. Sandow's extraordinary physical powers may be our own possession in but a faint degree; yet that they are that is an acquisition of no mean moment, for to what trained power we have we may some day owe our life. Is the argument without force as a plea for compulsory physical training?

Even in spite of himself, Mr. Sandow has become what is termed a professional athlete. To that fact, both in this country and in England, he doubtless owes much of his fame. But it is due to Mr. Sandow to say that he long resisted the clamour that he should exhibit his prowess for money and pursue professional gymnastics as a vocation. Not that, *per se*, the vocation is objectionable; but that, at the outset, he was under no compulsion to seek it as a profession, and was brought up in a rooted dislike to appear in public as a salaried exhibitor. The attraction to him was the enjoyment he took in

WRESTLING AND FEATS OF AGILITY AND STRENGTH

as an amateur. While indulging his tastes, as in an idle pastime, he broke, as will presently be seen, with his father, and that circumstance, coupled, possibly, with the fascinating glamour of the public arena, drew him at last into the profession. Like the high-minded and generous man he is, however, he cares little for the pecuniary rewards of his work. Had he wanted merely to make money, he would no doubt have taken to the pugilist's golden career ; but this, we know, was always abhorrent to him.

When, in process of time, Mr. Sandow came to know his own power, we can well imagine the pleasure he took in his continued muscular training. Modest as he is, and inclined, with the instincts of a gentleman, to repress rather than assert himself, we can hardly doubt that, at times, when he scores a great triumph, he takes honest pleasure in looking himself over with the lust of the eye and in the pride of life. With his magnificent physique, he would hardly be human if he did not. But his normal characteristic, professionally and privately, is self-effacement ; and though reliant and confident in his powers, he always bears himself modestly. Even when smarting from some provocation, or when a rival contestant unduly draws upon his courtesy and good-nature, he invariably places himself under rigid restraint. Only twice is he known, the occasions of which will afterwards be stated, to have departed from what, considering his strength, will be deemed a merciful habit.

Having regard to the interest of the subject, the earlier portion of the following pages will be confined to telling the tale of Mr. Sandow's public career. In chronicling the story, it is proper to say, that only indisputable facts will be set forth ; and it is the desire of Mr. Sandow, as it is that of the writer, that no exaggeration shall be indulged in and no colour given to the narrative beyond that which the facts themselves warrant. This, it is hoped, will be deemed to have been rigidly adhered to, in dealing with incidents which, in London especially, became the subject of much journalistic controversy in relation to Mr. Sandow and his defeated rivals. One thing may be said in this connection, and it is itself a guarantee of good faith, as well as good taste in the subject-character of the book, that Mr. Sandow has never made a claim for himself to which he had not a right, or which the facts themselves do not furnish the proof. After this fashion, and in the spirit we have indicated, we proceed, in the chapters which follow, to unfold the life-history which, with some misgiving as to our ability to do justice to the theme, we set out to relate.

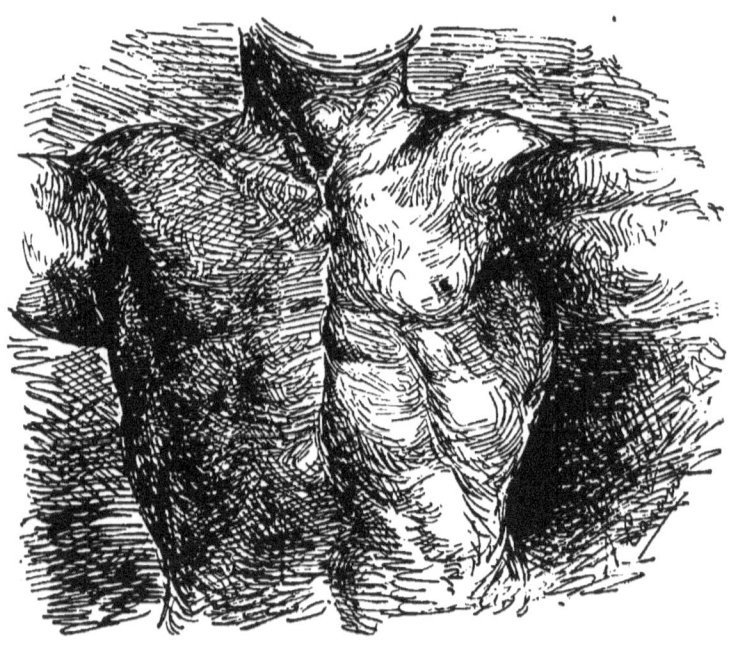

III.

SANDOW'S BOYHOOD AND EARLY LIFE.

A PERSONALITY so marked as that of Sandow, with such power latent in him, both of will and purpose, as would make of him the character he has become, presents, even in youth, many aspects of view, the presentation of which can hardly fail to be of interest to the reader. It is a trifle tedious, however, as most will admit, to dwell in minute detail on the early life of men who have subsequently made their mark in the world. We shall not fall into this error in treating of Mr. Sandow's youth-time, for all we might say would be to repeat the aphorism, doubtless in his case with variations, that "the child is father to the man." If we enlarged upon this topic, it would be to remark that while from his earliest years young Sandow had a fondness for athletics and exercised his muscles, even turtively when he was denied the opportunity to do so openly, he never dreamed of reaching the perfection of bodily development and muscular power he was afterwards to attain, or of becoming renowned on two hemispheres for mighty deeds of physical agility and strength. He had, nevertheless,

A BOY'S NATURAL AND HEALTHY DESIRE FOR DISTINCTION,

though, as yet, the field of his youthful tests of strength was a modest one, and immature were the powers which one day were to do great feats. In his ambition to train himself, he aimed at being thorough rather than showy. and, as he has counselled many a pupil in athletics since,

HE PUT HIS MIND INTO HIS EXERCISES.

The phrase, in Mr. Sandow's mouth, is worth dwelling upon, for, as he earnestly and persistently avows, it is the key to success as a gymnast. The difference is great, as every learner knows or ought to know, between going through certain exercises in a perfunctory and mechanical manner, and putting the muscles to the strain by concentrating the mind and will-power upon the manipulation of the weights, or whatever muscular exercise is being attempted. Exercise, he of course also maintains, should be systematic, persistent, and thorough. Without this, and disregarding his chief injunction, to put mind into your work, anything like proficiency cannot be reached. How assiduously and laboriously Sandow has himself trained, few men who have not some time or other equipped themselves for competitive contests can know. For years, as we have already observed, he did this for the love of it, and without thought that in the future he should turn his training into an arduous but profitable vocation. This fact, in telling the story of the athlete's early life, we may have occasion to repeat, for Mr. Sandow is fond of referring to it with his young pupils as an encouragement when they are apt to weary of assiduous exercise and the toil it entails. But this and other matters of practical interest we shall come to in the narrative of the life, upon which we now enter.

Eugene Sandow was born at Königsberg, Prussia, on the 2d of April, 1867. He is consequently now only in his twenty-seventh year. As a child he was healthy and well-formed, but there was nothing of the prodigy about him, physically or mentally ; nor were either of his parents of anything but normal physique. Up to his fifteenth year, indeed, young Eugene was of slight build and rather delicate constitution. His father, like all patriotic Germans, had served some five years in the army, but took to commerce as his life-vocation, and, in time, became a prosperous jeweller and dealer in precious stones and metals. This worthy citizen of Königsberg is now dead, as is his wife, Mr. Sandow's loving and devoted mother. A half-brother, who also is only of average physique, is a professor in the University of Gottingen. Sandow himself was an earnest student, and in his school-boy years was deemed a fair, all-round scholar, though he had a preference for mathematical studies, in which he was well versed and won honours. Contemporary with his college-days, he devoted himself with great ardour to all forms of gymnastic exercises and athletics. There he stood upon what was now to be commanding ground, for so successful was his training that he soon distinguished himself in all sports, and feats of agility and strength. In these he outrivalled even his senior schoolmates. He loved, beyond anything, to steal off to the gymnasium and the circus, and in the latter, with youthful but wayward ambition, longed to test his strength

The circus was, however, unhallowed ground with his good and honest parents, and, seeing their son drawn with uncovenanted bonds to the glittering arena, they put the place for him under interdict. This was a sore rebuff to young Eugene, but it led to the redoubling of his own home-efforts to become redoubtable as an athlete.

About this time young Sandow's holidays fell due, and his father, being in good circumstances and fond of his boy, who had been diligent in his studies, gave him the treat of taking him with him on a visit to Rome. Arriving at Rome, what the youthful scholar had imbibed of the classics led him to take keen interest in the art treasures of the Eternal City, particularly in the statuary, representing the gods and heroes of antiquity. Under the local influences of the place, his imagination repeopled the Corso and the Colosseum with the stalwart deities of Roman mythology and he seemed to see, as in a vision, the great pageant of a past day, with mighty concourses of people applauding their laurel-crowned favourites in the wrestler's arena. But, practically, he liked most to frequent the art-galleries, and there to hang about and admire the finely-sculptured figures of heathen deities and the

CHISELLED BEAUTY OF SOME HERCULEAN ATHLETE

or wrestler in the throes of a life or death struggle. With the inquiring mind of youth, he asked his father why our modern race had nothing to show in physical development like those lusty men of the olden time? Had the race deteriorated, or were the figures before him only the ideal creations of god-like men? His father's reply was a disappointment to him, for he had to admit that the race had suffered physical decline, and even in its choicest individual specimens had fallen grievously from its once mighty estate. Later ages, with their ignoble ideals, and the sordid habits and fashionable indulgences of the race, had wrought their due havoc—a havoc which the father took occasion to impress on the youth's mind, and the admonition was not lost. Eugene, contrasting his own slight figure with the mighty thews and graceful forms of the statued heroes about him, conceived the idea to train his body to the utmost pitch of perfection, and so approach, if he did not attain to, the

ANCIENT IDEAL OF PHYSICAL POWER AND BEAUTY.

Returning to his home, in the high ardour of emulation, he devoted himself, more assiduously than ever, to muscular training and the intelligent study of his frame, its capacities and functions. Every opportunity he took advantage of that seemed to further him in his work and brought him nearer the goal of his purpose. Many and furtive, at this time, were his resorts to the circus-tent and the wrestler's arena. But these were forbidden indulgences, and though he tried hard to give heed of his parent's injunctions, his ruling passion was often too strong for him. So all-impelling was his ambition at this period, that we find him repeatedly running away from home, and as repeatedly and ignominiously being brought back.

The battle was of long continuance between young Sandow's inclination and his duty to his parents. It ended at last in his going to the University of Gottingen, where, however, he was permitted a measure of indulgence in physical training. Winning his way, with the exception of the limitations imposed upon him, he pursued his academic studies with zeal and energy, which so commended him in the eyes of his parents that they permitted his proceeding to Brussels to study anatomy. This, it may be said in passing, was not the profession his parents had designed for him. The family were of the Lutheran faith, and its heads had hoped that Eugene, with his gifts and prospects, might take to the ministry. But in this they were disappointed, though they were not loth to see their son turn to the healing art as a profession. Ere long, however, they had a new disappointment, for Eugene, at the medical school, confined himself almost entirely to the anatomical course. Here the reader will, once more, perceive the undeviating bent of the young athlete's purpose. Yet most valuable, it must be said, was the intimate knowledge he gained of the structure and

MUSCLE-RAMIFICATION OF THE HUMAN FRAME.

It was the instruction he most needed in pursuing his training as an athlete, and almost beyond price has he since found the knowledge he then acquired. Meanwhile, it gave new stimulus and a fresh direction to his labours in muscle-culture, and brightened and widened the outlook on his cherished athletic pursuits.

Up to this time, though young Sandow had achieved no inconsiderable local fame as a skilled gymnast and wrestler, he had had no thought of indulging his tastes beyond the limits of the amateur. A quarrel just then with his father altered the condition of things with the young lad, and confronted him with a grave crisis in his life. His parent, seeking to curb Eugene's infatuation for his favourite pastime, cut off his money-allowance and threatened him with other embarrassing deprivations. The result did not meet the fond father's expectations. It threw the high-spirited lad on his own resources, and only too apparent were the resources available. The circus and the theatre became more than ever his resorts, and not unwilling, as we may well imagine, were his feet to go thither. Luck and his skill threw prize-money in his way, and now and then he earned a little by hiring himself out to sculptors and artists as a model.

Only precarious, however, was at this time young Eugene's means of livelihood, and soon he had seriously to debate with himself how or where else he could make money. In debating the question, he bethought himself of a quasi-professional visit to Holland. Before he left Brussels, Sandow made the acquaintance of a well-known and noted professor of athletics, who, at various periods and in different countries, was to figure in exhibitions with him.

SANDOW FIRST MEETS ATILLA.

This was Professor Atilla, who, at the time we are writing of, conducted a gymnastic training school at Brussels. Sandow's introduction to this expert instructor in physical education was due to the enthusiasm of some pupils of Atilla, who had caught sight of the young Prussian wrestler, entering a café opposite the gymnasium, while they were themselves receiving a lesson. Sandow was induced by his eager friends, who knew his skill, to bring himself to the knowledge of "the Professor" by exhibiting some of the more wonderful feats they had known him to perform. The exhibition proved an astonishment to Atilla, for he found that the youth not only surpassed all his pupils in dexterity and strength, but could do many things which the master was himself unable to perform. On the other hand, the partnership which grew out of this chance encounter was, while it lasted, of service to Sandow, for the latter learned something from the expert which was afterwards added to his own *repertoire*. Together, the two paid professional visits to Rotterdam, Antwerp, and other towns close by, and later on returned with the modest gains of their labour to Brussels. They also found at Leyden, among the students at the University, interested and well-paying pupils, to whom the athletes, for a time gave lessons.

IV.

SANDOW AS A STRONGMAN IN HOLLAND.

WITH no decided views as to where, after parting with Atilla, he would be likely to find employment, Sandow found the occasion urgent to go in search of it, for he was again entirely dependent upon his own resources. In passing from his native Prussia to Belgium, he left behind him not only those who knew and loved him, but, to some extent also, the interest actively felt throughout the Fatherland in wrestling and all manner of gymnastics. To the young adventurer the situation was more serious when he had to pass from Belgium into Holland, because this took him still further from hope of engagement, where he was known as an athlete, besides, as we have seen, having now to get along without his father's allowance. In proceeding to Amsterdam, he was venturing

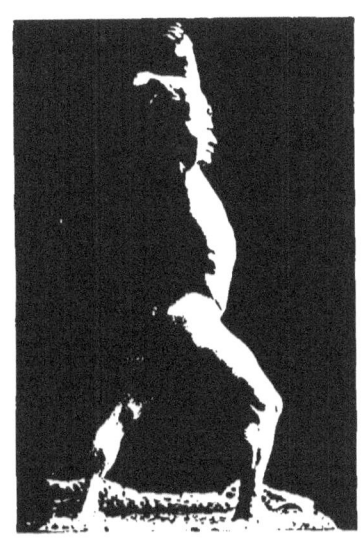

SANDOW. CLUB STUDIES. Sarony—Photo.

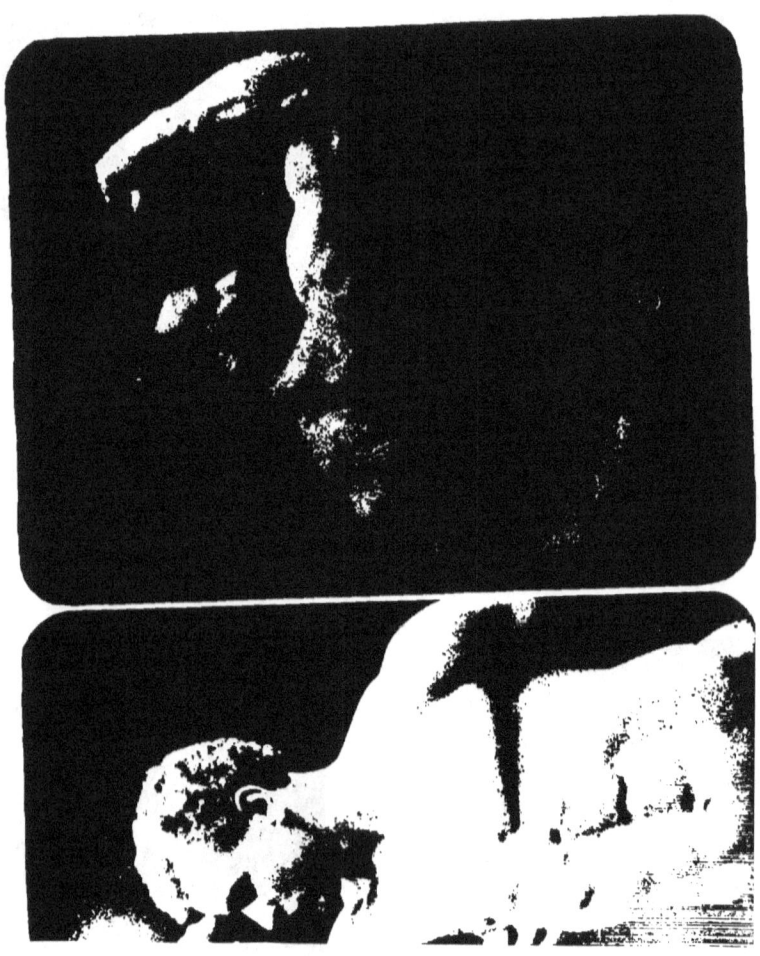

his barque on an entirely unknown sea. He as yet knew no
one in the city, though he possessed the pleasant manners and
frank, open countenance of one ere long certain to make
friends. He had, moreover, youth and hope on his side, and,
by this time, had acquired remarkable strength, with a varied
though miscellaneous experience of circuses, theatres, and
shows. At the chief theatres he sought employment as a
strongman, but strongman exhibitions, he was brusquely,
almost rudely told, were not then in vogue; while the manager of the "Paleis voor Volksvlyt" would not pay Sandow
the humble ten guilders ($4) a night the young athlete asked
for his services. At this juncture, when fortune most frowned,
his worthy father once more besought him to return home;
but, though without prospects, and in almost extreme need of
money, he refused. Depressed and crestfallen as he was,
with his hotel bill in arrears and not a little of his effects in
pawn, he yet had confidence in himself: in any case, he could
not brook the idea of acknowledging his life, so far, a failure.

ESCAPADE AT AMSTERDAM.

One day, when his store of money was quite gone, save a
mere pittance in his pocket, a daring scheme entered his head,
which, he thought, would be a novel mode, at least, of advertising himself, and might lead to his securing the employment which he now sorely needed. He was, as we have said,
unknown in Amsterdam, and had had no chance afforded him
to show his powers. What he did was to take a cabman into
confidence and arrange with him to drive him round the city
some morning between midnight and dawn. His purpose
was to visit all the weight-lifting machines scattered throughout the town, outside the closed cafés, and wreck each in
turn by a strong pull at the handle—a feat which only a very
powerful man like young Sandow could do. Dependent upon

the good-nature of the cabman, not only to keep his counsel as to what he intended to do, but for the necessary coin to put in the slot of each machine, he set out and only too well accomplished his purpose. In the morning, when the city was astir, every passer along the streets carried the news to the police stations, and soon bulletins were issued by the newspapers, saying that the city had been visited over night by a gang of ruffian marauders, who had, by their combined strength—so the account ran—dismantled and wrecked every weight-lifting machine. The whole city wondered at the deed, and for days it was the subject of universal talk. The authorities offered a thousand guilders reward for the discovery and capture of the miscreants. Every citizen, and of course every habited guardian of the city's nocturnal peace, had each his own theory of how the town came to be so invaded and the machines gutted. In time, the town breathed freely again ; the machines were repaired ; and the inexplicable deed was about forgotten. A second time, and, after a little, a third time, the city woke to a repetition of the machine-wrecking experience.

ARRESTED ; AMUSING SCENE AT THE POLICE STATION.

After the second of the wrecking exploits, it was of course not easy to guard against surprisal, for by this time the police were officiously on the *qui vive*, while every porter and night-watchman was but too anxious to obtain the civic reward. The cabman, with Sandow, had almost completed the third night's round when the latter was espied by a porter at one of the cafés just as he was giving the wrench to a machine which threw it out of gear and broke the springs. The porter, realizing the apparent strength of the nightly depredator, kept at a respectful distance from the strongman, but having the reward of the authorities before his eyes was not willing to lose the chance of bagging his game. Sandow, on the other hand, having sufficiently stirred up the city to interest in his nocturnal acts, was but too ready to reap his own peculiar reward and inwardly was not averse from arrest.

The porter, meanwhile, having rushed to the nearest police-office, brought with him a posse of constables, who collectively pounced upon young Sandow, who suffered himself to be taken to the station. There he was catechised by the sergeant-in-charge as to who were his confederates in crime, for no one imagined that the machines had been wrecked by a single pair of arms. Sandow's protestation that he alone did the deed was received at first as a joke. Again and again was he interrogated on the point and threatened with handcuffs and imprisonment. He, of course, continued to make but one answer, and as its possible truth began to dawn on the police they treated him with more politic consideration. At this, Sandow, with a nonchalant air, repeated his protest against arrest, for, as he naïvely observed, he had been merely exercising his arms, and in the slot of each machine had honestly paid the toll. Presently, a commissary of police appeared on the scene, and, with amazement and curiosity, heard Sandow's account of the affair and his demurral to the indignity of arrest. The comic aspect of the scene was reached when the culprit gave indisputable evidence on the biggest of the constables that he was the strong man he claimed to be, to the amusement of the inspector and the crowd that by this time had gathered in and about the police station.

After this amusing exhibition of strength, which quite won the heart of the old commissaire, Sandow was released on his own recognizances, promising to appear should action by the authorities be pressed, which, we may say here, was not the case. On the contrary, the young athlete became the lion of the town, and he and the cabman were escorted in triumph to the hotel where Sandow lodged, which has since become a great resort owing to its connection with the morning's incidents. There the entire staff of the establishment was for hours kept busy drawing beer for the enthusiastic populace that had followed Sandow and were talking in hilarious glee over the affair. A suite of fine rooms, in exchange for his previous humble domicile, was offered our hero by the hotel-proprietor, who had caught the contagion of excitement from the crowd and was eager to show his gratitude to Sandow for bringing him such welcome and unlooked-for custom. This custom, thanks to the now notorious athlete, was not evanescent, but grew daily in volume, especially while Sandow made the city his home; and the hotel-proprietor, it may be remarked, emphatically dates the founding of his fortune from the day on which the incident transpired which we have just related.

At the theatre, it may be added, which had refused Sandow a salary of ten guilders a night, he now obtained a prolonged engagement at twelve hundred guilders a week!

AT LONDON AND PARIS.

The success of the machine-wrecking hero at Amsterdam brought together again Atilla and his quondam partner and pupil. Together they resumed for a time their itinerant exhibitions and afterwards crossed over to London, where Atilla had secured an engagement at the Crystal Palace, Sydenham. There Atilla, shortly after appearing, had the ill-luck to meet with an accident on the stage, which terminated his engagement and threw both himself and Sandow out of employment. Shortly after this, Sandow drifted across to Paris, where dame Fortune again became fickle and for a while treated him churlishly. He made repeated but fruitless efforts to get an engagement, and failing in that became exigently hard up.

His ill-luck continued for some weeks, and only a forced resort to the pawn-shop enabled him to exist. To Sandow's surprise it was even difficult to hire himself out as a model. One day, after meeting with only mortifying rebuffs, the young athlete called upon a professor of anatomy, at the Academie des Beaux Arts. To the professor he made his usual request for employment and was met with the now familiar answer, that "just then he had no need of a model." Impatient at the stereotyped answer, he urged that he was a strong man and physically well-developed, adding, pathetically, that he would be thankful for even a day or two's engagement, that he might be fed. The professor, not heeding the appeal, or being in a hurry to get back to his class, turned to ascend the stair, leaving Sandow, in chagrin, to take himself off. But the latter was not thus to be got rid of, at least without giving the learned gentleman a practical proof of his strength. As the doctor, who was a large man, was mounting the stair, Sandow caught him by the legs, and with an easy, effortless movement he put him down at his side.

"*Mon Dieu*," said Esculapius, "you are indeed a phenomenon!"

"Yes," rejoined the athlete, "and if you give me a chance you will see what else I can do."

The doctor now invited Sandow to his class-room, where he exhibited his figure and some of his notable feats before an enthusiastic band of students, winning the deafening applause of all present, with a purse, to which each pupil contributed, containing two hundred francs. For several months, Sandow continued to exhibit at the Academy as a model, and also found remunerative work in giving private lessons as an athlete.

About this time, Sandow made the acquaintance of a strolling circus-man named Francois, with whom he made a lengthened tour with a pantomime show, Sandow contributing no little of the attraction by his gymnastic feats and unrivalled power as a wrestler. These exhibitions proving remunerative, Sandow finally embraced them as a profession, meeting henceforth an almost unbroken run of luck.

V.

SANDOW AS A WRESTLER IN ITALY.

THE tour in France with the pantomimist, François, was, as we have said, a remunerative one, and naturally so, for the play in which Sandow and his quondam partner appeared had many elements of interest and novelty. As a pantomime, it amused the people; while the combination of athlete and harlequin introduced a new feature in entertainments of the kind, which astonished as well as delighted the audiences that were drawn nightly to witness them. The pantomime was entitled *L'Afficheur* (the bill-sticker). It was composed as well as partly performed by Sandow and François, who appeared under the stage-designation of "les frères Rijos." Its original character may be apprehended from the circumstance that François, who personated harlequin, was dressed as a huge doll, which Sandow juggled and tossed about the stage, threw over walls, and pitched in at windows, with a freedom which for a time disguised from the audience the fact that it was a living man, and not a stage property, that was being shuttlecocked about. Amusingly labelled, harlequin was also thrown against walls, to which he clung, exhibiting, in ingeniously contrived changes of dress, the pictorial embellishments of the bill-sticker's art. The performance, though an amusing farce, gave opportunity for the display of Sandow's extraordinary dexterity and strength, and served well the purposes for which it had been temporarily taken up. From France Sandow and Francois passed on to Italy, where, at Rome, they met with continued success as they adapted the pantomime to the tastes and local circumstances of that country. With the company was an artist whom Sandow had known in Holland at the time of his machine-breaking escapade, and an evening was devoted to the giving of a benefit to this old confrere. To give eclat to the evening's performance, the artist begged Sandow to exhibit his prowess in some feats of strength other than those in which he was accustomed to appear. Anxious to favour his friend, he readily complied with the request, though he was without apparatus, which had to be borrowed or extemporised. After a little searching about, the necessary appurtenances were got together. Impressed into service, besides a set of dumb-bells, were a couple of pairs of railway-car wheels, with their axles; yet, with this motley apparatus, Mr. Sandow not only contributed his own share to the evening's entertainment, but achieved a triumph which threw into the shade the other performances of the occasion. So signal was his success, that the director of the local theatre called upon him to offer him a very liberal sum if he would abandon the pantomime and engage himself as a wrestler and performer of feats of strength. This he afterwards did, and won a name for himself in the Eternal City for his performances, which attracted King Humbert and the athlete-loving members of his court. He was, however, not unmindful of his partner, François, for he shared with him the profits of his exhibitions until he left the city.

While at Rome, Sandow had an opportunity of enhancing his fame as a wrestler, for in this capacity he had been giving lessons to the titled youth of the Italian court, as well as wining their admiration for his powers as an athlete. This came about in consequence of a challenge he received from Bartoletti, a notable Roman wrestler, not unknown in America, who offered to stake 5000 francs on the result of a wrestling match with Sandow. The challenge was accepted, and the issue was a surprise to all Rome, for Sandow's victory was an easy one and enabled him to put the prize-money in his pocket. On the morrow of the contest, the surprised winner was made the recipient, after the fashion of the South, of innumerable bouquets, with other souvenirs and presents, including many applications from those seeking to become pupils of Sandow in learning the art of athlete and wrestler. In a short time he had more pupils than he wanted, though the aggregate fees were welcomed in the improvement of his finances. From Rome, Sandow at length passed to Florence, whither his reputation had preceded him, and there he met with equal success, and was presented by the Athletic Club of the famous art city with a handsome gold medal in commemoration of his visit.

Subsequently, Sandow visited Venice and Milan, where he won further honours with the golden rewards of his work. At the latter city he received a new challenge from Bartoletti, who, it seems, was not satisfied that he had been fairly beaten, or was at least unwilling, without further trial of strength, to accept defeat. Sandow, good-naturedly humored the great wrestler, and the new match took place at the Theatre d'Alverne, with like results. Sandow again was victor. A new contestant at this time came upon the scene, named Sali, a man who was acknowledged to be the best wrestler in Italy, and had won repute in Australia, where he defeated every one of his opponents. The moment was an auspicious one for a trial of strength between an Italian and a Prussian, for Germany and Italy had just then joined the Triple Alliance, and the political movements of the time invested the match with an international importance. Sali, moreover, was known to be an ugly customer in a contest, a man who would do his utmost to beat his opponent, as well as to maintain the fame of his country. Public excitement rose to a high pitch over the match, and the gymnasium of Milan put up the money for the stakes. The day arranged for the contest came, but the sequel was not varied in Sali's case, though he stood well up to his work, and the match lasted over an hour. The honours once more fell to Sandow, who came off victor and received an enthusiastic ovation, with the usual accompaniment of presents of fruit, flowers, and bon-bons.

Subsequently, Sandow beat, in five minutes, Milo, a pupil of Sali's, and then proceeded to Venice, where he bought a villa, with the design of enjoying a brief vacation. Here he was induced, however, to forego his well-earned rest, and to issue a challenge, this time to any two wrestlers, whom he (Sandow) undertook to tackle at once, the stakes being 3000 francs. A number of would-be competitors came forward, attracted by the daring challenge ; but two only remained in the field to try their luck against the redoubtable Prussian. Their names were Sarini and Vocoli. Notwithstanding that the occasion was the first one in which Sandow had ventured to wrestle with two men at once, ten minutes sufficed for the contest, for within that brief space both athletes were on their backs.

Elated at his success, and being in admirably good form, Sandow now published a challenge, inviting three wrestlers to try their skill against him in one match ; the rules of the contest being that, as each man is successively thrown, he is considered out of the ring; though, until there is a fall, all may come upon the challenger at once. His three former opponents, Bartoletti, Sarini, and Vocoli, accepted the challenge, and, as may be imagined, great was the excitement which the proposed match evoked. It will appear tame, as well as the merest commonplace, to chronicle the result; which varied in no whit from that of the preceding matches, though Sandow had an unusually hard struggle of it to wrest victory from the grip of his triple opponents. The match lasted an hour and a half and was a triumph such as Sandow might well be proud of. Against professionals of so great repute, no wrestler has hitherto been known to contend,—in a match three against one,—and to come off conqueror.

So notable a contest could hardly be won without its being widely talked of and deservedly praised. For a time it was the chief topic in the German and Italian Press, and the theme of comment in all the Mediterranean cities and towns. It took place just at the period when the late German Emperor, Frederick William, was at San Remo under treatment for his throat—the malady which was soon to deprive the Fatherland of its loved monarch. Sandow's renown naturally reached the young king's ears, and Frederick sent a message to Venice commanding the presence at San Remo, of the Konigsberg athlete. The command, we need hardly say, was obeyed with eager alacrity, and Sandow had the honour of giving an exhibition of his prowess before the Kaiser and his royal consort. The Emperor, though sadly stricken now with his fatal malady, was himself a man of much strength, and naturally took pride in witnessing the feats which his athletic fellow-countryman had to show him. With an old-time pride in his own powers, Frederick took a complete pack of playing-cards and with a strong, quick turn of the wrists tore them in two. It might have been courtly etiquette to leave the Emperor to the enjoyment of the pride he felt in the work of his hands; but some one informed His Majesty that Sandow could beat him at his own trick, and it was with pleased surprise, and with no admixture of envy, that he witnessed *two* packs torn apart by the renowned athlete. After witnessing several other astonishing feats, the Kaiser took a ring of great value from his finger, which he had worn for eighteen years, and with frank heartiness presented it to Sandow, telling him, at the same time, that he was an honour to their common Fatherland, and that he could desire nothing more than that his army were made up of many such types of fine physical manhood. He added, with touching pathos, that, to possess Sandow's perfection of bodily health and strength, he would gladly exchange places with him, were it in his power to do so. He also expressed the hope that he might live to see Sandow his guest at Berlin. The ring, which he then placed on his subject's finger, is of beautiful French enamel, encircled with brilliants, with the initial F, and a crown over it composed of diamonds. Sandow naturally prizes it as the most cherished of his souvenir treasures.

After this memorable incident in the great athlete's career, Sandow returned to Venice, where he had an encounter with a wrestler, named Muller, whose unprofessional brutality in a match which ensued, gave Sandow occasion long to remember him with keenest dislike. He was, in truth, a terrible antagonist; being known to resort to infamous tactics—such as seeking to break his opponent's finger or limb—to get the better of his adversary and win a match, even through a foul. Sandow, though aware of Muller's vicious habit, was not loth to try odds with him, and the match was duly brought on, before an immense and highly wrought-up audience. Sandow entered the arena and confronted his adversary with his usual pluck and coolness. After some minutes' struggling and a few feints, Muller saw that he was not likely to throw his opponent and he then attempted to play his old game, which

Sandow, for the time being, foiled. Muller persisted, however, in his tactics, and endeavoured to get Sandow at a disadvantage, seizing hold of wrist, arm or limb, in turn, with the evident design of breaking or disabling it. But Sandow was wary, and for a further while succeeded in checkmating his purpose, until, with both hands, Muller fastened upon Sandow's right forearm and tried to snap it at the wrist, and at the same time, with a supreme effort, he forced two fingers of his right hand an inch deep into Sandow's flesh, crushing the veins till they burst, and causing him intense pain. This dastardly act, so foully committed, caused Sandow, for the first and only time in his life, when wrestling, to lose his temper, though not his presence of mind. With all his strength, Sandow, by an alert movement, jerked back his left arm, and, closing upon Muller, threw both arms round his body, between the waist and the chest, and drew his opponent towards him in a very bear's-hug until Muller's face blackened and blood gushed from his mouth, and he fell upon the floor as if he were dead. The defence was the act of a minute; but it left Muller, not dead, but with four broken ribs, from which it took him a long time to recover; while Sandow was disabled for four months, the veins being torn in his arm, and the nerve-fibres greatly lacerated. Even to-day, though five years have passed, Sandow speaks of the circumstance with keenest regret and no little reticence; though it was Muller's own perfidy that provoked Sandow to administer the merited, but unrestrained chastisement.

It was at Venice, shortly after his recovery, that Sandow made the acquaintance of the English artist, Aubrey Hunt, R. A., whose admiration of the fine physical development of the great athlete led him to paint the now well-known picture of Sandow in the Coliseum at Rome, in the character of a gladiator. It was from this artist that Sandow first heard of Sampson's nightly challenges at the Westminster Aquarium to any athlete who would come forward and do the feats performed by himself or his pupil, Cyclops. On the evening of the day Sandow was apprised of the challenge, he was already on his way to London, with what results—if the reader is not yet aware of them—the next chapter will disclose.

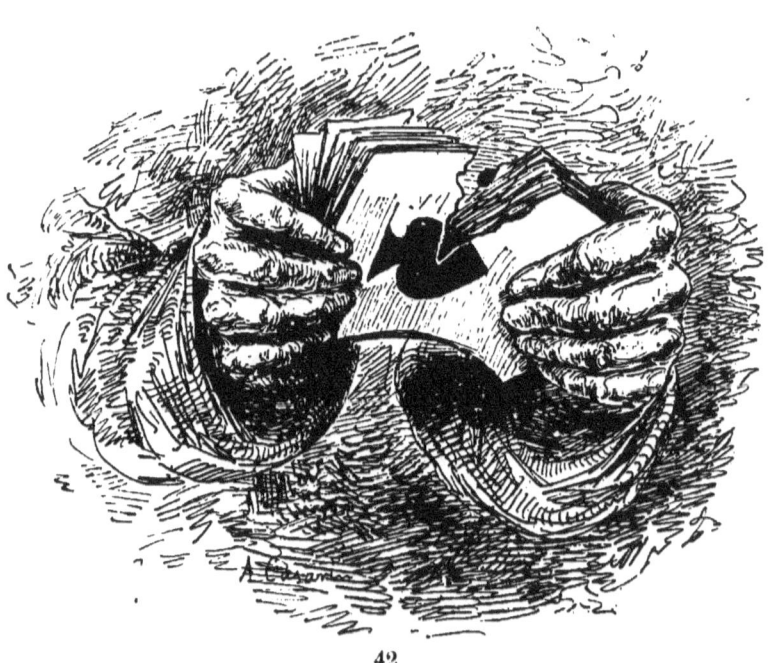

VI.

SANDOW WINS HIS FIRST LAURELS IN LONDON.

SANDOW was in his twenty-third year when he came to London, attracted, as we have seen, by the rather braggart challenges of Samson on behalf of himself and his pupil, Cyclops. Within a couple of days after his arrival, the young Prussian athlete became the subject of as much public talk as if he were some royal personage whom the clubs, the privileged class and society in general had agreed to treat as the lion of the season. This was due to Sandow's immediate and enthusiastically-hailed triumph over Samson's protege, including the winning of a £100 stake at the Royal Aquarium, Westminster, after a contest of unprecedented excitement and thrilling interest. Something, no doubt, was due, also, to the suddenness with which the then unknown strongman had alighted upon the world's metropolis and won so signal a victory, to the modest yet confident demeanour of the victor—in marked contrast to the manner and bearing of his challenger —and, especially, to the interest excited in the classic beauty and fine physical development of the newcomer's form and person. London, we know, loves dearly to have an idol, however brief and inconstant—if we are to take the cynic's word for it—may be its idolatry. In Sandow's coming on the scene, the great city was, however, justified, in the matter both of physique and prowess, in its *penchant* for idols. Here is the

PEN-PORTRAIT OF THE YOUNG ATHLETE,

as he then appeared, taken by a representative of the London *Daily Telegraph*, and published in that eminent journal, November 4th, 1889. "Personally he (Sandow) is a short, but perfectly-built young man of twenty-two years of age, with a face of somewhat ancient Greek type, but with the clear blue eyes and curling fair hair of the Teuton. When in evening dress there is nothing specially remarkable about this quiet-mannered, good-natured youth; but when he takes off his coat and prepares for action, the extraordinary development of the arms, shoulders and back muscles is marvellously striking. It is no exaggeration to say that the statue of the 'Farnese Hercules' (see illustration) is not more powerfully modelled; the muscles stand out under a clear white skin in high relief, and suggest the gnarled roots of old trees."

Similar testimony to Sandow's attractiveness of person and rare physical development appeared, we may say, in the entire metropolitan press; and, for months, almost every English journal gave columns to the chronicling of Sandow's wonderful doings. So great and wide-reaching was the interest taken in him, that, throughout the British islands, the worship of muscle became a cult, and every phase of athleticism, with reminiscences of those who had notably figured in them, was minutely and unweariedly discussed. Referring to our hero's achievements, and their genuine and legitimate character, another London journal (the *Morning Advertiser*) at the period admiringly remarked, that "there was no doubt about the extraordinary performance of the victor (Sandow) in lifting and holding at arm's length a full-grown man, or in toying with a 150 lb. bar of iron as if it were an average dumbbell. Unadorned efforts of this sort," the same journal goes on to say, in allusion to the stage-feats of other strongmen exhibitors, "speak for themselves, and appeal far more effectively to the admiring astonishment of the beholder than the snapping of chains or the bending and twisting of metal rods —exhibitions which many people observe with a haunting distrust in their perfect authenticity, and a feeling that there is more of trick or 'knack' than of downright physical prowess in them."

Sandow came to London, however, to win a wager, not specially to be written about or merely looked at. As we have said, he had heard of Samson's challenge, on behalf of Cyclops, and he came to see, and if possible rival, the feats of this strongman and pocket the stake which Samson nightly made a show of putting up. For some weeks, Samson had been giving exhibitions in the London Aquarium, under the boastful designation of "the strongest man on earth," and lately he had associated with himself a pupil, whom he styled Cyclops. We owe to a London journal the following description of master and pupil. "Samson," says this authority, "who is about thirty years old, was born at Baden (other biographers speak of him as an Alsatian), and for a long time he has travelled through Europe and America, exhibiting feats of strength, breaking thick chains with blows of his wrist, and twisting stout steel ropes by the mere muscular expansion of his chest and his arms. From the pictures of him which are exhibited throughout London one would be apt to think that he is a giant in proportions and formidable in appearance. Decked in war-paint he certainly looks a personage undesirable to tackle, but seen in his every-day garments faultlessly cut and of superlative fit—he might easily be taken for a debonair attache in Her Majesty's diplomatic service, more accustomed to dance attendance at levees than to work for chainsmiths by breaking steel links across his breast. Cyclops, who when at home rejoices in the humble patronymic of Franz, is nineteen years old, and hails from the good old town of Hamburg, where he worked as a blacksmith, until he came over to England to earn fame and £10 a week as Samson's pupil. In stature he is far beyond his master. Huge in frame, fat and bull-necked, with a good-humoured, expressionless face, he appears to have found the exact vocation nature designed him for, in lifting huge weights above his head and swinging ponderous dumb-bells around his body as if they were children's toys."

These were the two men against whom Sandow had come to London to pit himself, and in entering the lists against them he was to bring both them and himself into fame. On the evening of the day on which Sandow and his friend Atilla reached London, they duly presented themselves at the Westminster Aquarium and took note of the feats ostentatiously performed by Samson and his pupil. So easy of accomplishment did the whole performance appear to Sandow, that he was with difficulty restrained from at once taking up the evening's challenge. Next night found Atilla and Sandow again in their places, accompanied by the latter's agent, Mr. Albert Fleming. When Samson appeared, and, as was his wont, offered a £100 note to any one present who should do the feats of strength about to be performed by Cyclops, Atilla took up the gage of battle for Sandow, who himself presently came upon the stage, and, as a local chronicler has it, naively asked if the money, were it won by an outsider, would be paid over on the spot. The wary young athlete was soon assured on this point. For at this juncture the £100 note was produced by Samson and, amid the applause of an expectant multitude, was handed to the chairman of the Aquarium Company. This gentleman was Captain Molesworth, who sat in a private box, near by the stage, and agreed to act as referee.

Sandow's unexpected appearance on the platform was a surprise to Samson, who had grown accustomed to make his nightly challenge without fear of loss to his pocket, though the youth's presence had a decidedly stimulating effect on the audience. This effect was increased when it became evident that the newcomer was no novice in the strongman's art, and could do, not only the feats Cyclops was wont nightly to perform, but rival him in the more difficult tasks his master, under pressure of the situation, had set him to do. But this will be best gathered from a detailed account of what occurred, though in planning how best to furnish this we were confronted with an embarrassing dilemma. Our first thought was to tell the story, as modestly as possible, in our own words. The evening's incidents, however, were so exciting, and led to so much altercation, that, on reflection, we decided to abandon our purpose and let another tell the tale, who would not be suspected of partizanship, and in whose dispassionate judgment the reader might have confidence. If our own fairness can be relied upon, we would venture to say, that in selecting the report of the London *Sportsman* (see issue of Oct. 30, 1889), we have drawn not only upon an admittedly high authority, but upon a journal whose account is distinguished, over that of many of its compeers, by truthfulness and moderation. We add that, for obvious reasons, we give the report entire, and not any garbled extracts from it.

"FEATS OF STRENGTH AT THE AQUARIUM—SAMSON'S PUPIL DEFEATED.

"SAMSON, who has been drawing excellent houses in the theatre at the Aquarium, had an extra good attendance last evening. Somehow a rumor, circulated in sporting circles, led to the conclusion that the performance of 'the strongest man in the world' and his pupil would be embellished by an unrehearsed effort worthy of the attention of all amateurs of feats of strength. Samson has been issuing challenges nightly, offering sums of money to any one who would undertake to perform the same feats of strength as his pupil, Cyclops, who lifts dumb-bells and heavy weights with comparative ease. The fame of Cyclops has spread, and the offers made by Samson have apparently not fallen on a deaf ear, for last evening, at the commencement of the performance, an amateur in evening dress, presented by a friend, took up the gage, and, divesting himself of his upper garments, stood out the beau ideal of an athlete. Herr Eugene Sandow, a young amateur from Königsberg, in Pomerania, of twenty-two years of age, a friend of Professor Atilla, well known in Germany, France, and England, for the successes he has obtained in his particular line of business, professed his willingness to imitate the feats performed by the pupil of Samson. Herr Sandow had come expressly from Venice, not to detract from the performance which has been carried on so successfully at the Aquarium, but to prove what a strong man is, and to take up the defiance which has been issued. He is an immensely powerful young fellow, weighing 14 stone, 6 lb., with a chest measurement of 45½ inches, something enormous for his age. His muscles stand out like iron bands, and those who saw him when he removed his dress-coat and vest felt certain that Cyclops would find a foeman worthy of his steel. Herr Sandow has never before competed in public, but as an amateur he has won fame in Italy, Holland, Belgium, Russia, Switzerland, Austria, and France. In the first-named country he met three of the best wrestlers, and let them all come on at the same time, treating them as some modern Samson, and vanquished them easily. The tutor of Cyclops determined that the novice should have no easy task, and after posting the hundred pounds in the hands of Captain Molesworth, the manager of the Aquarium, he varied the programme so as to tax not only the strength of his pupil, but that of Herr Sandow, to the very utmost.

"Cyclops, notwithstanding the disapprobation manifested by the audience at the departure from the usual programme, took up 50 lbs. with his left hand and then lifted up another hundredweight with his right, elevating his arm until it was raised above his head. His opponent stepped to the weights, and amid loud applause carried the hundredweight twice above his head, outrivalling his predecessor. The next performance was with the heavy dumb-bells, and this being successfully done by the novice, some of the audience pretended that he had won. Cyclops, lying on the ground, raised the heavier dumb-bell at arm's length, and this feat was repeated amid renewed applause by Herr Sandow. Samson did not appear to relish the outlook, and instead of the ordinary block of stone, which weighs 400 lbs,. and has to be raised with one finger, extra weights were put on it, until Cyclops fairly staggered beneath the load. An outcry was raised at this further departure from the programme, and some of the audience exclaimed that the feat had been performed for the first time. There was a good deal of quibbling as to whether the challenge thrown out by Samson really meant that any one accepting should be compelled to outrival Cyclops in feats of strength, or merely implied that the usual performance should be gone through. When silence was restored, Herr Sandow implied his willingness to do what his opponent had done, and he was cheered to the echo when he repeated the feat, which was accomplished with far more ease by him. Cries and counter-cries were heard, and a soldier made himself conspicuous in the gallery by the animated manner in which he took the part of the newcomer, and by taunting Samson with having lost his money. Samson, with difficulty, managed to obtain a hearing, explaining that the hundred pounds he had offered could only be claimed by a man who could do all that Cyclops had done, and not what had been accomplished on any previous night. He said that he had ten more tricks for his pupil to perform, and that they must be successfully imitated by any one pretending to the money deposited in the hands of the stakeholder. This version of the offer made at the commencement of the proceedings was not accepted by the majority of the spectators, who were of opinion that the newcomer had done all, if not more, than had been required of him. The scene became more and more animated, and Herr Sandow sat down to rest while each party strove in vain to get a hearing; and Samson's attempts to address the public were met by cries of 'No more performances'! 'No more tricks'! 'Part!' The tumult was stilled when Captain Molesworth, who was seated in one of the private boxes, asked for a hearing, saying that he was stakeholder, and that the public might accept that statement as a guarantee that he would see fair play, and only give the money when it was won.

"Samson failed to relish the comments of the audience and lost his temper, while the life-guardsman in the gallery called on the amateur not to attempt any other performance. In vain 'the strongest man' strove to prolong the agony, but the more he did so, the more hostile became the audience, and finally Mr. Frank Hinde, who evidently has the ear of the *habitues* of the Aquarium, went to the rescue, obtaining silence for Captain Molesworth, who said that he had decided that thus far the amateur had fairly accomplished all that had been asked of him. As referee he called on Samson to name any two feats which he considered were the best to prove the superiority of his pupil, intimating his intention of handing the £100 over to Herr Sandow should he succeed in successfully imitating them. After another scene, Cyclops' *impresario* consented that the two feats should be named, although he argued that the last performance of his pupil had not been gone through by the man who had taken up his challenge. The first trial of strength proposed was imitated with apparent ease by Herr Sandow, and then Cyclops, girding up his loins for a final attempt, staggered under the heavy 150 lb. dumb-bell, while, with a hundredweight in the left hand, he elevated his arm twice above his head, letting the weight fall with a thud on the stage. Shouts of 'don't do it; Don't try it; you have already won your money!' greeted the young Königsberger as he stepped forward and felt the weight of the heavy dumb-bell. He smiled in response to the warnings, and poising the ponderous bell in the right hand he grasped the other weight, and, bending his left arm, slowly raised and lowered it, not twice but seven times, amid thunders of applause.

"Samson again lost his temper, but there was no appeal against the decision of the referee, who had the hundred pounds handed over. He attempted to persuade the public that his challenge had been misunderstood, but his explanations were laughed to scorn, and even Professor Atilla, who mounted the platform, failed to obtain a hearing. Samson again offered £100 to Herr Sandow if on Saturday night next, on the same stage and place, he would perform all the feats attempted by Cyclops, and again the challenge was accepted. The young Königsberger very wisely listened to the advice of his friends, and refrained from taking up the gage thrown down by Samson, who defied him then and there to go through his particular performance of chain-snapping, breaking wire cables, etc., but he professed his willingness to demolish two steel chains with his naked fist if Samson would give the £500 he had been accustomed to offer any one who could imitate the feat he accomplishes nightly with his gloved hand. An explanation was volunteered that the £500 would become the property of Herr Sandow if on Saturday night he would perform the same feats as the challenger. This offer was accepted; the wire strands were examined as Samson burst them asunder by inflating his chest, and also the chains, which were snapped by a violent effort of the muscles of his right arm. The great trial of strength will therefore take place in the theatre of the Aquarium on Saturday evening next. Samson has not disguised his intention of struggling to maintain his reputation as 'the strongest man in the world,' and the meeting will be an exciting one, although Herr Sandow has set himself a task in undertaking to outrival master and pupil in one and the same performance."

Such, in detail, are the incidents of the evening's lively competition; yet, severe as the test was, the honours were unquestionably Sandow's. After Samson's exhibition of petulance on the stage, it will little surprise the reader to learn that that redoubtable angrily repudiated his discomfiture. His pupil, after Sandow, figuratively speaking, had put his Cyclopean eye out, is related to have "burst into tears and wept like a child." Neither of these things, however, detracts from the fairness of the young Königsberger's victory or alters the emphatic results of the challenge. Samson, naturally, had many admirers, drawn to him by his feats and long engagement at the Aquarium, and, of course, it was possible for him, through his friends, to try to turn defeat into a victory and create sympathy for himself by posing as the victim of the judge's decision. Like his Biblical prototype, he set fire to the foxes' tails of his pupil's and his own discomfiture, and sent them running through the Sandow-Philistines' corn. All this little helped the defeated strongman, however, except as he shared later in the results of the increased attractions of the Aquarium, consequent upon the coming of Sandow and his embarrassing acceptance of Samson's challenge. This, indeed, was no slight aid, for it would be difficult to overstate the furor throughout London occasioned by the public shearing of Samson, by the male-Delilah who had so dauntlessly appeared on the scene."

Into almost every nook and corner of the great city had news entered of the battle of the giants, and public excitement rose to fever-heat in anticipation of the greater contest, to be settled on the following Saturday evening. The subject, for the time, indeed dwarfed every other topic of local and even international importance, including the Parnell Commission, then sitting ; while the Press found in it prolific themes of interesting comment and, among the journalistic wits, amusing reflection. One of the latter, a writer in the Glasgow *Herald*, made sport of the affair by affecting to bewail the public loss of its most cherished allusions—the taking upon trust the claim of the idol showman to be bigger, smaller, or fatter than the rest of us—which turns out, after test, to be disenchantment. "It is painful to hear," observes the writer, "that Samson, 'the strongest man on earth,' has been subjected to destructive criticism, as if he were an historical myth like William Tell's apple, Richard the Third's hump, or Cambronne's defiance at Waterloo. . . . The appearance of a second strongest man on earth—or the equal at least of the strongest man's second—on the same stage as the first is disheartening, and it seems not improbable that any number of these superlatives may be forthcoming. A young man, who as far as physique is concerned was not to be compared with either Samson or his pupil, Mr. Frank Cyclops, seems to have lifted weights and performed other feats with a facility which astonished and delighted the audience. The stranger also won a £100 note, which had been boldly wagered by Samson in support of his declaration that nobody was strong enough to earn it. The £100 note was a financial detail. The stranger who won it from M. Samson seems to have earned it fairly and squarely, and he now probably appreciates the theory recently discovered by a distinguished prelate that betting is not a sheer loss of money—to the man who wins. The principal part of the whole business is the loss of another of our illusions. M. Samson, who appears to hail from Alsace, ought not to have walked into a trap of his own setting. It did not matter what country gave him birth so long as he was a permanent ideal to those people who never expect to handle 500 lb. weights to look up to.

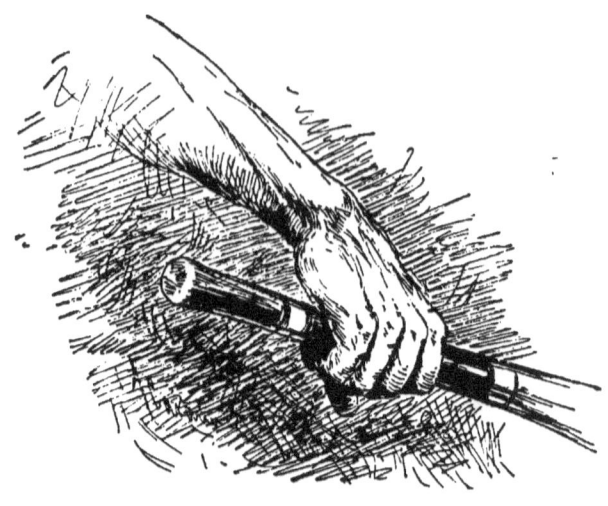

VII.

DEFEATS SAMSON AT THE WESTMINSTER AQUARIUM.

The evening of Saturday, Nov. 2d, 1889, proved another red-letter night in Sandow's phenomenal career. The contest, as we have seen, had been eagerly looked forward to in almost all circles in the metropolis, certainly in all circles interested in athletic sports. Hardly anything, indeed, could more emphasize the love for athletics in the English nation than the interest manifested in this contest, between Sandow and Samson. "If the fate of the Empire," observes a London journalist of the period, "had hung in the balance, more keenness in the coming match could not have been shown." Looking back now upon it, it is no doubt to be regretted that the results of the encounter were not more satisfactory; though it may fairly be asserted that Mr. Sandow, at least, was in no way responsible for that. The terms of the challenge were admittedly loose, and the conditions, if they were drawn up at all, can hardly be said to have been acted upon when the contest came off. It may matter little now ; but the unbiased reader will be apt to say, as very probably was said at the time, that if the comparative strength of the two contestants was worth determining, it should have been determined conclusively and on a proper basis. There should have been no room either for shuffling or display of temper ; while both athletes ought to have been tried by the same tests, and the genuineness of the tests vouched for beyond suspicion or peradventure. To omit the guarantees of good faith in a contest of such moment, was to discredit whatever was legitimate in the performance. Nor were the issues of the contest helped when the challenger, hugging his resentment, refused to ' give a lead ' or to attempt the feats performed by his opponent. Under the circumstances, it is not surprising that Fate and the audience of the evening went against him.

But we are anticipating, and, moreover, in danger of giving a colour to the story we have to relate, which we would fain avoid. Lest we should do so, we will, with the reader's permission, follow the plan adopted in the previous chapter, of giving an account of the evening's incidents as supplied by the contemporary press, selecting, as dispassionately as possible, a narrative which shall tell the facts as they happened, without prejudice or exaggeration. To vary the representation of the London journals quoted in these pages, we will, on the present occasion, draw upon the *Daily News*' report, which we append as follows:—

" STRONG MEN IN RIVALRY—AN UPROARIOUS NIGHT AT THE AQUARIUM.

" Athletics had an exciting, not to say uproarious, field-night on Saturday at the Westminster Aquarium. The beauty of the turnstile system was well illustrated, for without these revolving barriers of iron the eager multitude would probably have carried the place by storm. The rival athletes, Samson, the Alsatian, and Sandow, the German, gave a public trial of strength, with the object of proving which was 'the strongest man on earth.' That was the promised bill of fare. The theatre was crowded, many of the seats having, according to rumour, fetched sums varying from one to five pounds. With two or three somewhat conspicuous examples the audience was one of men—gentlemen, many of them, of position (as for example, Lord Bury, who sat in one of the stalls), shining lights from the Pelican Club, sporting men from the Stock Exchange, besides Mr. John Hollingshead, Mr. Edgar Bruce, Mr. Kyrle Bellew, Lieutenant Dan Godfrey, and Colonel North. The tobacco smoke, gradually rising like incense on high, became thick enough to dissect before the curtain rose; but you could make out quite clearly that the theatre was packed with a very fair specimen of athletic humanity, men who could give a literally striking account of themselves in a scrimmage. Samson came to the footlights, dapper, radiant in medals, tights, and dainty boots, and smiling with confidence. He made a little speech, the first of an unconscionable series delivered or attempted before the business ended. He wanted fair play; he offered £500 to any one who would come on the stage and perform the feats he performed. Never mind where such a man came from; let him appear. There was no response, only a babel of cries from the audience. By and by a gentleman, not unknown in the prize ring, advanced to the footlights, stretched forth his hand, and said, 'Sandow is not far off. He is in a room.' Meanwhile Samson, after a considerable pause, made signs of beginning; but upon cries for 'Sandow,' put on his cloak, and strutted backwards and forwards on the boards. He again came forward to say that he did not want the challenge money; he would give it to any hospital; and there was a very pretty hubbub thus early in pit and gallery. With a fine flourish, the Samsonian cloak was now assumed the wearer explaining that there was too much draught; he did not want to kill himself; he would retire to abide events in his dressing-room. The next commotion was caused by a number of gentlemen reaching the stage by flying leaps from a side-box, sweating and touzled after fighting their way through a frenzied mob in the crowded hall outside.

"At last Sandow entered, amidst general cheering; Captain Molesworth, apologizing for the delay caused by the besieged state of the building and its approaches, announced that the Marquis of Queensberry and Lord de Clifford had consented to act as judges, and asked for fair play for the competitors. The two men were in the centre of the stage. Samson, in his gay athletic costume, Sandow in a plain, pink, sleeveless under-vest, and black trousers encircled by a leather belt. Neither of the men is of more than medium height, but their arms were a rare spectacle, by reason of masses of muscle brought by practice to the hardness of metal. Sandow, however, has the more spacious chest and largest arms, and a connoisseur would probably fancy him as best for a trial of sheer strength. We had soon to hear the programme, as explained by Captain Molesworth. This was to be a continuation of a trial of the previous Tuesday, when Samson had to pay £100 won from his pupil, Cyclops, to Sandow, and when Samson offered £500 if Sandow could perform his feats. Upon this statement being made, Samson came forward to protest, and an interval of uproar ensued. Captain Molesworth begged the audience to hear Samson, who was of Southern temperament and excitable. Thus adjured, the crowd were silent until Samson insisted that his challenge was for £500. against the same sum. If that amount was not forthcoming, let it be £100 against £100. Another swell of clamour followed, Samson excitedly declaring, 'If he wins, he takes my name. I leave the stage. If he loses, I give the money to an hospital.' Captain Molesworth said that he should be sorry to see the audience disappointed, and therefore he undertook, in the name of the Royal Aquarium Company, that £100 should be placed against Samson's £500. Crowding to the front, Samson now insisted that he offered nothing of the kind, asked for fair play, and remarked that he would be taken for a fool to offer £500 against £100. At last it was settled that it was to be £100 against £100, and in course of time Samson threw down a number of lengths of what appeared to be iron gas-piping, and left them to carry on further disputation with the judges. Sandow stood back, his tremendous arms folded upon his broad chest, his clean-cut head, covered with short, close curls, held straight upon a Titanic neck; altogether a model for the statue that he seemed to be.

"A beginning might never have been made but for the judges, who decided that Samson must do the feats of strength he was in the habit of doing every night. This decided action on the part of the Marquis of Queensberry and Lord de Clifford evoked from a well-known wit in the front stalls the remark, 'Ah, I always said the House of Lords was a useful institution.' The first feat was with one of the iron pipes. Samson belaboured himself upon the chest, leg, and arms, bending it and straightening it back again by the blows. He did it gracefully and swiftly. Sandow laboured more, was clumsy, and took more time, but he performed the feat. After the inevitable discussion raised by Samson on the stage and a tumult amongst the audience, who, apparently, were by this time largely on the side of the phlegmatic German, Captain Molesworth stated that Sandow wished it to be known that he had never done the trick in his life before. Samson darted to the front and dashed into a speech that was drowned in uproar, save the one sentence, 'Why, he did it six years ago.' Next came a prolonged squabble about a strap trick, which the judges decided Sandow was not called upon to imitate. Then there was a feat of breaking a wire rope fastened round the chest. Samson performed it with the neatness of one accustomed to the trick of twisting the ends of the wire strands together. Sandow was obviously unacquainted with the knack, and it was only after prompting from the audience as to the twisting, and several fruitless efforts, that he succeeded. It was a splendid effort of strength. The man seemed like to burst in his effort to obtain the requisite expansion of chest, and when the iron rope burst asunder, like the withes of the Philistines around the limbs of the original Samson, the audience leaped to their feet, and shook the place with deafening cheers. The next thing was a contention and uproar about some bottle trick of Samson's not sanctioned by the judges; then chain bracelets were brought forward. Samson, always theatrical, put one set on his forearm, and offered one to Sandow. It was too small for such an arm, and he rejected it with a slight gesture of contempt. There seemed to be a hitch here. As the chain ring which fitted Samson would not go on Sandow's arm, how would the House of Lords get out of the difficulty? For once Sandow abandoned his statuesque attitude. To the astonishment of all he whipped out from his trouser-pocket an armlet of his own, and it was then necessary to wait until the audience had bawled themselves out. The proof of equality with an emergency was another feather in Sandow's cap. The unfortunate Samson protested, gesticulated, argued, trod the deck, and generally cavorted around. Another appeal to the House of Lords was a matter of course. At last the rivals put on their chains and smashed them by sheer expansion of muscle, the one as cleverly as the other. Samson snatched up the fragments of Sandow's armlet and ran about shaking them derisively, asserting that they were not of the same material. A gentleman in the audience, however, handed up an invoice from a Leicester Square firm certifying that they had supplied to Sandow one dozen yards of jack chain, the same as used by Samson. It was a long while before order could be restored, the incident apparently being regarded by the audience as a clincher.

"Enter Samson a little later, to hurl a heap of chain upon the stage, and shout 'I give him £1,000 if he breaks it.' Furious yells rent the smoky atmosphere. Samson donned the everlasting toga, and palaver the hundred-and-twentieth, or thereabouts, reminded the judges that their post was no sinecure. It was as good as a game to note the contrast between the quicksilvery Alsatian here, there, and everywhere, and the stolid German, with folded arms and lips closed like a trap, standing a motionless sentinel in the background. By and by Samson broke a piece off a chain, Captain Molesworth (not by any means for the first time during the evening) interceding with the audience not to disturb him by their interruptions. Samson himself shouted 'I have not had fair play at all,' an ill-timed remark which filled the cup of disfavour to the brim. There were at this time many demands from the gallery for a trial of lifting weights, but no notice was taken of them. Other propositions were made, amidst much talkee talkee on the stage, without avail. Samson's cloak was now off, and now on, and a more than usually tiresome consultation was ended by Captain Molesworth stating that the judges had decided that as Mr. Samson would not give a lead, Mr. Sandow might perform some feats of his own. The young German accordingly lifted a stiffened and upright man from the ground, and performed some astonishing feats with a Brobdignagian dumb-bell, weighing 150 lbs. Some of the feats Samson, from the side of the stage, applauded as heartily as any one; but he raised another hurricane by the prelude of an attempted speech. 'I give Mr. Sandow credit,' he said, 'for his strength. I knew Mr. Sandow a long time——' The audience effectually prevented the conclusion of the sentence. Sandow went on with his feats, and Samson made a rush at a chain which he and his assistants were manipulating, and tried to drag it away. Foiled in the attempt, he rushed about the stage shouting 'It is his own material.' The tempest was not allayed when a gentleman in the stall offered Samson £50 if he would do what Sandow had done with the dumb-bell, and Sandow's manager publicly challenged him to the same test. The challenge was not accepted. Midnight was by this time approaching, and Captain Molesworth virtually closed the programme by announcing, amidst general cheering, that the judges had decided that Sandow had done everything that Samson had done. The audience gave the victorious man an ovation, and it was then observed that Samson had disappeared from the stage. Special cheers were given for the judges and for Captain Molesworth, and there were calls for the rival gladiators to publicly shake hands. Samson, however, was seen no more; but Sandow, in a few words of German, returned thanks."

The honours of this second public trial of strength, the reader will agree with us, were, beyond question, again Sandow's. We repeat, it is a pity that the results of the contest were not more satisfactory; though it cannot be said that the aggrieved Samson was in any sense wronged or failed to receive British fair play. Throughout the evening's performance the injured manner and irate mood of the man were much against him; while it was easy to see, from the contrast in the bearing of his opponent, as well as from his unquestioned prowess, why the audience was demonstratively in the latter's favour. Another point of obvious regret was the absence of a well-arranged and agreed-upon programme, and of the sureties, as to the *bona fides* of the feats to be severally performed, which ought to have been provided for in the preliminaries. As matters turned out, Sandow, though he undoubtedly won Samson's challenge, was not paid over the wager (nor has it been paid to this day), while the latter carried his wrangling from the stage to the Press, and, for a time, made what capital he could by posing in public as a martyr. On this point, and on that of the relative merits of the two athletes' feats of strength, let us quote a provincial journal, which, two days after the contest, published the following sensible view of the affair. Says the Birmingham *Gazette* (Nov. 5, 1889): "Samson is still unsatisfied. He is too good a sportsman not to acknowledge that he has met a formidable antagonist, but he declares that he has not been allowed to put him to the test for which he, Samson, stipulated. The point which the public with collective common-sense has seized upon, is that Samson has been proclaiming himself 'the strongest man on earth.' Now the question arises; How can that matter be determined? Is it by lifting weights, by breaking chains and rods, by any of the various forms of physical endurance—such as standing on one leg, or by any other device or means that we can put the matter to proof? Samson insists upon the contests being those more or less tricky ones with which he has made the public familiar. But these do not satisfy the public, who much prefer the simpler kinds of tests, such as weight-lifting and dumb-bell exercises.

"Upon the latter basis Sandow has unquestionably won the

challenge fairly. It seems a pity that the difficulties cannot be solved by some such plan as the independent establishment of a series of tests by a competent authority, to which both men shall submit. Nothing creates so unpleasant a feeling among the British public as any suspicion of unfair play, and Samson may rely upon it that public spirit will support him in every effort he makes to obtain a fair trial. Something more will be wanted, however, than his suggested feat of snapping a chain. Obviously that is much too tricky a performance to be accepted as a test of strength."

That Mr. Sandow was only too well aware of the unsatisfactory issues of his match with Mr. Samson, and eager to put the strength of each competitor to a proper test, is clear from his ready acceptance of a new challenge, which Samson had issued on the night of the contest. Samson's challenge was affixed to a rather rambling setting-forth of his so-called grievances, and to a correction, which he seemed to think necessary, of misstatements (*sic*) respecting the late match appearing in the Press. Explicit, as well as reasonable, as are the terms of the proposed match which Mr. Sandow, through his agent, expressed his willingness to consent to, nothing came of it. Here, however, is the letter, which was addressed to the editor of the *Sportsman* and published in the issue of that journal for Nov. 6th.

"*To the Editor of the Sportsman*, Sir : With reference to the challenge published in your paper on Monday to the effect that Mr. Samson is willing to stake £5,000, to Mr. Sandow's £500, if Mr. Sandow can break on his arm chains to be produced by Mr. Samson, I, on behalf of Mr. Sandow (the vanquisher of Cyclops and Samson), wish to say that he will accept this challenge on the following conditions :—1. That the chains shall be selected by a jury of three gentlemen, to be named by the editor of any influential newspaper. 2. That the chains selected shall be made to fit the arm of each competitor, shall be bought, packed, and sealed by the jury, the seals to be broken on the stage in front of the audience. 3. That before the contest takes place all financial transactions connected with the competition of last Saturday night shall be satisfactorily settled, according to the decision of the judges on that occasion. 4. That in the event of neither Mr. Sandow nor Mr. Samson succeeding in breaking the chains so produced on the stage before the jury, then Mr. Sandow shall name six *bona fide* feats of strength which he is prepared to go through, and if Mr. Samson succeeds in doing three of these feats, then Mr. Sandow will repay him the £500, which he won from him on Saturday night.

"Yours, etc.,
"ALBERT FLEMING,
"*Manager for Mr. Sandow.*
"LONDON, *Nov.* 5."

Though nothing, as we have said, came of this further match, interest in the competitors by no means flagged. Great, indeed, was the strongman "boom" at the Aquarium, where Mr. Samson continued his now well-advertised performances, and especially at the Alhambra, which had secured Mr. Sandow for a lengthened engagement. The drawing qualities of the latter were, we need hardly say, most conspicuous, including not only all athletic and would-be athletic London, but royalty, also, and the flower of the nobility, *plus* the élite of Mayfair and Belgravia. Royalty was represented by H. R. H. the Prince of Wales and H. R. H. the Duke of Edinburgh, both of whom, we learn from the Press of the period, were interested spectators at Mr. Sandow's exhibitions. The former paid Mr. Sandow the compliment of a visit to his robing-room and asked to possess his photograph, while the latter was equally enthusiastic and exchanged gifts with the renowned athlete.

The Press, it is pleasant to state, maintained unabated interest in the performances of the young Prussian athlete, and ably seconded his efforts to disseminate intelligent views on the subject of physical education and the perfect development of the human form. This was an important service, as Mr. Sandow was now devoting his leisure, in addition to his Alhambra duties, to giving lessons on muscular training. Gratifying, consequently, was this appreciation on the part of the Press, for it will be understood that Mr. Sandow considers his exhibitions a very minor though onerous function, in comparison with the interest he feels in athletics in their relation to bodily health and the physical equipment of the race. One of the newspaper comments to which we have referred, and in which the great athlete heartily coincided, was the remark of an editor, given currency to in his journal, to the effect, "that it was singular that while a fine physique is constantly proving to be as much an object of admiration as it ever was in England, the simple and easy means of securing that advantage to every man, in degree, should not be more generally cultivated." The remark applies with like force to the New World, and is pertinent to the condition of things that obtain here, in the neglected matter of physical training.

VIII.

SANDOW IN SCOTLAND AND AT THE CENTRES OF INDUSTRIAL ENGLAND.

In the Press, the "War of the Titans," as the journalists loved to call the Samson-Sandow contest, was still being fought over and stimulated, to an extraordinary degree, the great London market for the display of Herculean wares. Sandow's exhibitions at the Alhambra, in which he was professionally assisted by his old friend, M. Atilla, continued nightly to crowd the house to the doors, and to extort unbounded, even extravagant, applause from the audience. The fame of the lionised athlete also brought him within the region of art, comic as well as serious, for Sandow's name was in the mouth of all London, and his feats were made material for travesty by the illustrated Press, in connection with the Parliamentary chiefs of the time and their respective sayings and doings. Perhaps the most amusing of these burlesques was the St. Stephen's Review cartoon, which represented Mr. Gladstone and Mr. Balfour, in athlete garb, rivalling each other, in elevating over their heads heavy dumb-bells labelled the "Irish Question." But serious art did not withhold its tribute, for Sandow was made an artistic study of in photography at the atelier of Mr. Van der Weyde, who desired, as he expressed it, to place before the public "a living Greek statue," taking the splendidly developed athlete as his model. The painting by Mr. Aubrey Hunt, R.A., representing Mr. Sandow as a gladiator in an arena at Rome, was a further tribute to the young Königsberger's fame. The well-poised small head, close curls, broad shoulders, and sinewy arms are displayed to capital advantage in this striking and finely-painted picture from Mr. Hunt's easel. In it, the great athlete stands, lightly clad in a tiger-skin and sandals, in the centre of an immense arena, the indistinct mass of gaily-dressed spectators forming an artistic background to the picture.

Mr. Sandow was now to make his bow to the athlete-loving people of the English Provinces, and there, for a brief space, let us follow him, for the fashion of London was to become the vogue also in the great centres of England's industries. Before setting out, however, let us record the incidents of an evening's exhibition at the Alhambra, for, so far, we have seen the invincible in competition only with his would-be rivals. To the *Sporting Life*, of Nov. 19th, we are indebted for the following introductory and chronicle:

"SANDOW AT THE ALHAMBRA.

"There are few things," says the *Sporting Life* reporter, "which excite an Englishman's admiration more than an act that requires a deal of nerve in its fulfilment. Any act of intrepidity, daring, or physical strength will elicit unstinted applause from the average

Briton, whose boast is that his games are open to all comers, neither country nor colour barred, and although he is beaten oft-times, all his opponents receive fair play. It will be fresh in the minds of our readers how the modern Samson offered £100 to any one who would perform the feats of strength performed by his pupil Cyclops, and £500 to any one who could perform the feats that he had been showing daily at the Royal Aquarium, Westminster. How unexpectedly, one evening a sturdy young fellow lightly stepped on the stage, accepted the offer of £100, wielded the heavy weights about as if they were playthings, and claimed the offered 'century.' Having vanquished the pupil, he volunteered to beat the master, and this he also accomplished in such a quiet and unassuming manner that the conqueror of Samson (as Professor Atilla delights in calling him) cannot but fail to command admiration as a man of extraordinary strength and physical development. In all countries and at all times there have been men of great strength, some of them possessing a muscular power so far beyond belief that one cannot help thinking that some exaggeration must have cropped up in the records handed down to us of their doings. But after seeing a display of bodily strength similar to the display given last evening by Sandow at the Alhambra, one becomes reconciled to the doings of these wonderful athletes. To revert to the doings of Sandow and Atilla last night. Directly the number was hoisted the audience commenced clapping. Professor Atilla was the first to occupy the stage, wielding 56 lb. weights and holding 112 lbs. up with one arm. Then he wielded a bar of steel weighing 90 lbs., and finished with balancing it on his chin. After wielding a dumb-bell of 150 lbs. he bent backward over a chair, and, returning, brought the 150 lb. dumb-bell with him—a very creditable feat of strength which the audience applauded. Then Herr Sandow tripped lightly on the stage, attired in pink tights, a blue vest, and his breast covered with medals. His coming was greeted with prolonged cheers. He commenced his entertainment by posing and then, putting both hands at the back of his head, moved his biceps in a marvellous manner. Catching up the 150 lb. dumb-bell, he moves slowly and gracefully with it, apparently without an effort, turns somersaults, and makes a mere plaything of this dumb-bell. Then, by way of varying the entertainment, he lifts an attendant, weighing 10 stone, about from side to side, and wonderfully holds him up above his head with one arm. He then picks up a larger dumb-bell, weighing 300 lbs. and raises it up twice with one hand. Sitting down on the stage, a board is rested on his knees and shoulders, and every available weight on the stage is placed upon it. As the last straw, Professor Atilla jumped on with a club, and the curtain fell amidst tremendous cheers. Upon being recalled, the Professor and Sandow playfully threw the 150 lb. bell backwards and forwards to each other and retired, but were again recalled before the curtain. C. A. Samson and Cyclops were interested spectators, but Sandow's performances are purely feats of strength. He neither breaks chains nor wires, but confines himself to weight-lifting only—the entertainment being a most marvellous exhibition."

As a pendant to this, we may be suffered to quote a personal portrait of Mr. Sandow, from a Liverpool paper of a little later period. "The refined manner," says the report, " in which Sandow goes through his performance is not understood except by those who have seen him. In appearance, the athlete is not by any means the ponderous being that is imagined. There is a conspicuous absence of the brutal proportions supposed to accompany muscular power, Sandow possessing one of the most symmetrical figures it is possible for the developed male to be endowed with. He is positively handsome in form, feature, face, and limb, the only proportion appearing somewhat out of balance being the enormous muscular development from shoulder to wrist, his arms seeming to have been hewn out of marble. The athlete's manner, moreover, is gracious and pleasing. He is the beau-ideal of athletic elegance ; he is not a big man, being of average size, though lithe and rapid in action and movement. Nor is there any painful exertion in his manipulations : on the contrary, he maintains a serene, calm, and easy demeanour throughout his arduous performance."

Mr. Sandow's tour of the Provinces, accompanied by Professor Atilla, extended from February till May, 1890, and covered visits to the following and other towns—Bristol, Bradford, Nottingham, Manchester, Sheffield, Leeds, Halifax, Huddersfield, Boston, Preston, Liverpool, Hull, Newcastle, York, Chester, Lancaster, Rochdale, and Derby. Everywhere a hearty reception awaited the now famous athlete, who astonished as well as delighted his audiences by his deft skill and prodigious strength. In some of the larger cities, such as Birmingham, Manchester, and Liverpool, the interest in athletics was manifestly quickened by the exhibition of Mr. Sandow's performances, and especially by the private exhibitions he was called upon to give to medical men and local athletes, who marvelled at the Prussian strongman's "mountains of muscle" and phenomenal strength.

The spring of 1891 Mr. Sandow also passed in paying successful professional visits to Birmingham and Liverpool. At both these cities the renowned athlete created great excitement and roused to a high pitch public interest in athletics. For the period of his sojourn in Birmingham, the Winter Gardens, where he gave his exhibitions, were crowded nightly by immense audiences, and the same is to be said of Mr. Sandow's appearances at Hengler's Circus, Liverpool. At each of these cities the Athletic Clubs vied with each other in paying courtesy to their distinguished guest, while the medical profession, in both cities also, made the great athlete the subject of admiring critical examination. During these *seances* with the medicos, Mr. Sandow good-naturedly gave demonstrations of his wonderful powers, including the lifting of men, over 16 stone in weight, from the ground at arm's length on to a table, and the tearing in two, by the strength of the wrists, of one pack, and on another occasion of two packs, of playing cards.

The Christmas holiday season of 1892 found Sandow, by special invitation, at the Scottish capital, giving exhibitions of his strength at a Carnival held in the Waverley Market,

Edinburgh. There "monster gatherings, numbering as many as 20,000 people, greeted the great athlete with Scottish heartiness and ardour. Nor were his admirers those only who saw him at the Carnival ; on the streets of the fair city, if we may trust the local chroniclers of the Press, he was followed by crowds, who " appeared to derive the liveliest satisfaction from observing all his movements. The amount of interest," remarks the Edinburgh *Evening Dispatch*, "his performances have aroused extends far beyond the ordinary Carnival audience. Many people have gone expressly to see him who never before honoured the Carnival with their presence, and his astounding feats have been the subject of universal comment in Edinburgh society for the past ten days." The *Scotsman*, the great Liberal organ of the Northern capital, was equally complimentary in its greeting of Sandow, as was the press of Glasgow, the sister city of the West, when the strongman paid it a visit. At Glasgow, Sandow's performances were hailed with the same fervour, and immense audiences filled the "Gaiety" and the "Scotia," where he successively appeared, to witness the unique and artistic display of muscle. Here, also, the medical faculty turned out in force to inspect and admire the champion strongman's physical frame. "Quiz," "The Bailie," the North British *Daily Mail*, and the Glasgow *Evening News* devoted columns to the chronicling of Sandow's feats—one of these journals noting the fact that some of the good people of the city, actuated by conscientious scruples, were prevented from witnessing Sandow's prowess, in consequence of his exhibitions taking place in an uncovenanted hall. Says the journal in question :—"The amount of interest aroused in medical as well as social circles has extended all over the world, and many more would be highly interested and become admirers of his wonderful ability, if it were not for the prudish spirit against being within the walls of a music-hall ! Of his performances, however, nothing but praise can be given."

IX.

WITH GOLIATH AT THE ROYAL MUSIC HALL, HOLBORN.

In the autumn of 1890,—to revert to the doings of that year—Mr. Sandow returned to London after his successful tour in the North. He had now become a very familiar figure and a great favourite with the frequenters of the theatres and variety-entertainment haunts in the metropolis. He began the season with an engagement at the Royal Music Hall, Holborn, with a programme which in its drawing qualities eclipsed all previous attractions and made him anew the sensation of the year. The great athlete had been, manifestly, increasing his strength and still further developing his wonderful muscular powers. At any time, it was a pleasure to witness his exhibitions, for, as a performer, he gave universal delight by the unaffected way in which he got through even his most difficult tasks, avoiding the poses, grimaces, and stage swaggerings, with which professional strongmen are too apt to decorate their feats. In addition to the extended *repertoire* Sandow had now to offer for the entertainment and wonder of his nightly audiences, he had brought with him to "The Holborn" a veritable giant, whom he had picked up doing the work of a stone-quarryman near Aix-la-Chapelle. This phenomenon was named "Goliath," and could hardly have been dwarfed by his namesake of Gath, champion of the Philistines, who measured, we are told, "six cubits and a span." Such a massive and rough-hewn block of muscular humanity probably never appeared on the stage before. He is 6 ft. 2½ inches in height, and weighs 27 stone ! "Goliath," observes a London reporter in attempting a description of this stage giant, "is of fearful and wonderful uncomeliness: he has hands big enough to let him use pillow-cases in daily wear as gloves. His measurements round chest, arm and head are phenomenal. At present he has not been educated to many stage tricks, and limits his share in the performance to walking round with a cannon weighing 400 lbs. on his shoulder. Sandow, however, makes up for the monotony of his partner's show by some really marvellous feats of strength, including the lifting of Goliath from the ground with one finger, and poising him overhead with one hand."

Of this man of almost fabled proportions, we shall get a fuller description, as well as an account of Sandow's new exhibition, in another source—that of the *Sunday Times* (Sep. 20, 1890)—which we herewith introduce to the reader :

"THE TWO GIANTS.

"As I am standing on the stage of the Royal Music Hall, chatting with Captain Taylor, the courteous manager, a young man, clad in a dark tweed suit, with a buff waistcoat, emerges from the wing, and stands, cigarette in mouth, watching the motions of the stage carpenters setting the stage. Captain Taylor introduces him as Mr. Sandow. The abnormal muscular development which makes him unique among living men is hidden in his street attire, and in his face, or in what is visible of his figure, there is nothing to speak of his extraordinary strength. The face and figure both look a little boyish. After a minute's chat on indifferent subjects he invites me to his dressing-room, on a level with the stage, in which the paraphernalia used in his performance are kept. In the corner is his "dumb-bell," two huge masses of metal united by a steel bar, and weighing in all 312 lbs., 12 lbs. heavier than that he used at the Alhambra, and much more difficult of manipulation, owing to its increase of several inches in length. This he invites me to examine. With considerable difficulty I manage to support it, staggering under its weight, when he insinuates a casual forefinger about the bar and relieves me of the burden. Various other of the weights which figure in his performance are standing about the room, and as he chats with me he performs, in an easy manner, various feats with them, and ends by getting me to stand on the palm of one of his hands while he lifts me on to the dressing-table. He dissipates the wonder of this performance by telling me that he is going to do the same with Goliath, the new giant, who scales twenty-seven stone. 'I am expecting him every minute,' he says. 'Come back to the stage, perhaps he is here now.' We go back, and there, sure enough, stands Goliath, a huge mountain of flesh and bone, standing well over six feet, with a chest measurement of Heaven knows how many inches, and huge face like a pantomime mask. This gentleman's hand measures over twelve inches from the tip of the thumb to that of the little finger, and the silver ring on the index of his right hand slides easily, with room to spare, over any two of my fingers. His hat covers my head and rests upon my shoulders. He bestrides the narrow stage like a Colossus, and Sandow, standing beside him, is a mere pigmy, though he is almost as much Goliath's superior in mere brute force as he is in deftness. Goliath speaks no English, but has a fashion of expressing friendly interest in anything going forward by a sort of short grunt, which shakes the building.

"Three stage-carpenters are now arranging upon the platform in the centre of the stage the tools with which the two giants are to perform their nightly work. The great dumb-bell, the smaller article of the same kind, the hundred-weight and half-hundred-weights, and similar trifles are symmetrically placed about the carpet-covered daïs, and Sandow, leisurely stripping off his coat, proceeds to rehearsal. It is a mere 'music-cue' rehearsal, and much the most interesting part of the performance is gone through in dumb-show. It transpires during its progress that the gigantic Goliath has very little to do except to pose as a foil for his infinitely stronger and cleverer companion. Sandow's penultimate performance is really marvellous: Goliath, girt by a leathern band, stands upon a raised platform which brings his waist about on a level with his companion's elbow; in the easiest manner possible, Sandow puts his hand under the belt and walks off with his huge companion held at full length. Perhaps the most remarkable of the feats performed by the latter is that known as the 'Roman Column.' A pole of burnished steel, some twelve feet in height, is made fast to the stage by cords and guys. Halfway up are two stout cross-bars, each projecting six inches in length, and from the summit hang two steel chains, ended by hooks of the same metal. These fit into rings affixed by straps to Sandow's legs a little below the knee. With his feet upon the cross-bars, and unsupported save by the chains, he bends the upper part of his body backwards and downwards until his extended hands touch the stage. On the stage lies one of the big dumb-bells, weighing 150 lbs. This he grasps, and with a terrific effort, which makes the muscles of his arms, legs and loins start out like lianas on a forest tree, draws it up higher and higher, till his body is at right angles with the steel pole, and the dumb-bell is held triumphantly at arm's length above his head. The performance ends by Sandow making a bridge of his body upon the stage, supporting the body, chest upwards, with his arms and knees. A board, pierced with three holes, one of which encircles his neck, while the other two fit about his knees, is put upon him, and on this the whole of his paraphernalia, supplemented by the weights of the three stage carpenters and the gigantic Goliath, is piled, Sandow supporting the whole weight, a total of 2,400 lb.—900 lbs. more than he supported last year at the Alhambra."

To these feats in the Sandow-cum-Goliath performance others were added as the exhibition drew still larger crowds and won greater fervour of nightly applause. These included the lifting, while lying on his back on the stage, of the 312 lb. dumbbell with two men seated upon it—a weight of some 620 lbs. Another startling feat, performed by Sandow, was the swinging round and round of a dumb-bell weighing 150 lbs. with two attendants suspended therefrom. The giant Goliath then makes his appearance, carrying a cannon, weighing 400 lbs., on his shoulders; after which Goliath stands in a square open frame, and Sandow from the top lifts him with one finger from the stage. Prolonged was the cheering which nightly followed this marvellous exhibition of human strength. To realize, adequately, what this astonishing feat is, the reader must remember Goliath's enormous weight, of 27 stone: his chest measurement is 65 inches, and his height 6 ft. 2½ inches. The contrast between the two men—Sandow and the Westphalian—is sharp in the extreme. Goliath is huge, lumbering, and unprepossessing; Sandow medium-sized, agile, and a model of compactness and symmetry. "From head to heel," as the *Newcastle Chronicle* has described him, "there is not a bad point in him. His features are of a bold classical type; his head is well-shaped and balanced upon a white and muscular neck; his shoulders are immensely broad; and in every limb— from mighty arm to shapely calf—the muscles stand out firm and rounded as bosses of steel."

Sandow's next engagement was at the London Pavilion, where, having parted with the ogre of Music Hall notoriety, he appeared with a promising phenomenon of muscularity, christened Loris. "Loris,"—we quote from a contemporary journal, the *Evening Post and News*,—"so far plays with such trifles as 56 lb. and 90 lb. weights, and does not essay a bigger dumb-bell than one reckoned at 140 lbs.; but he handles these with the utmost ease, and as much grace as is compatible with severe muscular effort. Sandow's display has been so often described that it is unnecessary to comment on it. He gets through his work with as little appearance of excessive effort as need be, and about both young men there is a pleasant absence of the theatrical swagger of many performers in the same line of business."

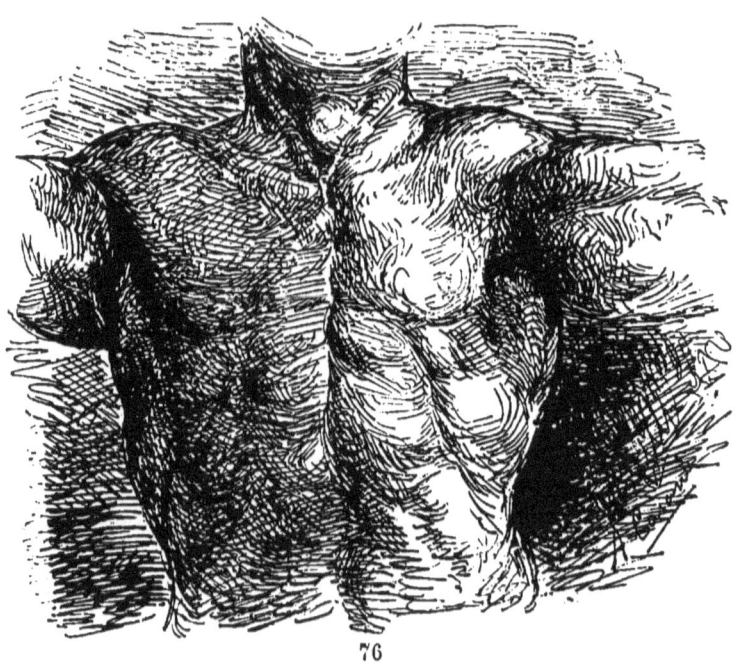

X.
ANOTHER STRONGMAN CONTEST.

WHILE Sandow was still exhibiting at the Holborn Music Hall, there was talk of another trial of strength among strongmen. The match, on this occasion, was to be between Sandow and one of two brothers, named McCann, professionally known as "Hercules" and "Samson."* These brothers were Englishmen—Birmingham men, we believe,—then under engagement at the Tivoli Theatre, London. The one to be pitted against Sandow was known as "Hercules," or, *en famille*, Henry McCann. The twin strongmen, it appears, entertained doubts as to the weight accuracy of Sandow's 312 lb. dumb-bell, which he was wont to raise nightly at his performances, and to

* This was not the Alsatian of that stage-name, who styles himself "the strongest man on earth." He, since his defeat by Sandow, added a " p " to his name, and now calls himself "Sampson."—*Ed.*

put the matter to test they offered to stake £50 if he (Sandow) "is able to lift a weight of 250 lb. avoirdupois with one hand from the shoulder to arm's length above the head"—a feat they (the McCann Brothers) deemed well-nigh impossible, frankly admitting, at any rate, that they could not do the feat.

This proposal was, however, but a preliminary skirmish, which at length, after interesting discussion in the columns of *The Era* and *The Star*, developed into a well-arranged and accepted challenge, covering not only the point above raised, but a threefold trial of strength, on each side, the stakes being £100 a side, with £50 additional to try conclusions in the lifting of the 250 lb. weight. The terms of the match were agreed to by both parties, and the respective stakes were deposited at the office of *The Sportsman*, the match to take place on the afternoon of Dec. 10th, 1890.

That the sequel of this match unfortunately brought a miscarriage of justice, is a matter the writer of this, for obvious reasons,—chiefly those of good taste,—does not desire to dwell upon. He contents himself with saying that, as will presently be seen, Sandow performed *four* out of the six feats set down on the programme, while Hercules performed but *three*, and failed entirely to attempt the specific feats Sandow had put forward for his opponent's test. If there is doubt at all of the injustice of the issue, we fail to find support for it in four-fifths of the reports of the contest published in the London newspapers of the period. With all but unanimity of voice the journals condemn the verdict.

Under the circumstances, it would be more than unseemly were we to give our own version of the contest. Happily, we need not here depart from the procedure we have heretofore acted upon, in allowing a contemporary English journal, of high repute, to furnish a report of the match. The following is from the *Morning Post*, Dec. 11, 1890 :

THE MATCH WITH "HERCULES" M'CANN.

"In fulfilment of an agreement entered into between the well-known strong men, Eugene Sandow and the Brothers McCann, professionally known as Hercules and Samson, a weight-lifting competition took place yesterday afternoon (10th Dec., 1890), on the stage of the Royal Music Hall, Holborn. Much interest was taken in the contest, which was witnessed by a large gathering of spectators, occupying all parts of the house. The competition consisted of six genuine feats of strength, three to be selected by Sandow, and three by one of the Brothers McCann, the feats to be named on the day of the contest, and the stakes to be £100 a side. In addition, the Brothers McCann offered Sandow the sum of £50, if he should succeed in lifting a weight of 250 lbs. with one hand, from the shoulder at arm's length above the head. The trial of strength was apart from the competition proper, and rendered Sandow liable for £50 in the event of failure. Sandow also agreed to give the Brothers McCann £50, win or lose, in consideration of their competing at the Royal Music Hall, where he is at present engaged. The performance was announced for 3 P. M., and after some delay, occasioned by the testing of the weights, a formality elaborately carried out upon two weighing machines, the curtain was raised, and disclosed Sandow and Hercules ready to engage in competition. The Marquis of Queensberry, Professor Atkinson, and Mr. Shirley B. Jevons, who officiated as judges, occupied seats on the platform, as did many supporters of both athletes. The preliminaries briefly disposed of, Sandow proceeded to take up the challenge to lift the 250 lb. weight for £50 The young German performed the feat—perhaps the most difficult in the programme—with complete success, and was loudly applauded. The regular contest then began, Hercules setting the first task, which was to raise with the left hand from the ground at arm's length above the head a weight of 170 lbs. The challenger accomplished the feat, and Sandow was also successful at the third attempt, the limit allowed for each trial. Sandow then, amidst renewed applause, raised a dumb-bell weighing 226 lbs. with his right hand at arm's length above the head. Hercules declined to attempt the feat, his decision provoking loud cries of disapprobation and a good deal of hissing. He then proceeded to his own test, which was to raise with the left hand at arm's length above the head a weight of 155 lbs. This he accomplished satisfactorily, as did Sandow, who, like his opponent, raised the dumb-bell twice.

The second of Sandow's tests was to lift a weight of 198 lbs. with the left hand at arm's length above the head. The challenger, however, could not quite succeed in straightening his arm, and gave up at the second attempt. Hercules, therefore, was not called upon, and was thus spared a great tax on his strength. The last of the three tests set by Hercules was to raise simultaneously two dumb-bells straight from the ground at arm's length above the head, the weight for the right hand being 120 lbs. and that for the left 112 lbs. This feat the challenger performed with apparent ease at the first attempt. Somewhat to the surprise of his supporters, Sandow was unequal to the task in which his opponent's superior weight was obviously an advantage. The last of the six trials was initiated by Sandow. It consisted in raising at arm's length above the head 210 lbs. with the right hand and 49 lbs. with the left. This very trying feat was accomplished after two unsuccessful attempts, and called forth a general burst of cheering. There were loud and prolonged cries for McCann, but, as before, Hercules refused the challenge, heedless of the ironical remarks showered upon him. This brought the contest to a close. The net result being that while Sandow had performed four out of the six feats, two of his own and two of his opponents, Hercules had accomplished only three—his own, having declined to attempt two of the tests set by Sandow, and being under no obligation to try the third. The judges then retired to draw up their decision, which was considered by the vast majority of the spectators to be almost a certainty for Sandow, whose splendid proportions and modest bearing, coupled with the fact that he had undergone far greater exertion than his opponent, made him a strong favourite. After an absence of a quarter of an hour, the judges returned, and the Marquis of Queensberry announced that Hercules had won the competition, while Sandow had gained the special prize of £50 already referred to. The decision came as a complete surprise, and was received with an outburst of dissent from all parts of the building. A scene of wild excitement followed, and in the general din, Sandow, who attempted to speak, could not obtain a hearing. At last there came a temporary lull, and a man, who proved to be Sandow's old rival, the Alsatian Samson, pushed his way to the front of the platform and declared, amidst tremendous cheering, that Sandow was the winner. This was evidently the popular verdict, the decision of the judges being incomprehensible to most of those present. The curtain was then lowered, and the spectators dispersed."

We may be permitted one further word bearing on the extraordinary and inexplicable issue of this contest. We have said that the judges' decision (which according to the terms set forth in the articles was to be final) was received with amazement and dissatisfaction. That no other result could follow the announcement of such a judgment, will be apparent by reiterating and briefly analyzing the facts. The articles of agreement say that the competition shall consist of six feats, three to be selected by each side. How were these competitive tests severally performed? Hercules set and did his own three feats, to which, inferentially, he had been habituated. Sandow successfully performed two of these, though, unaccustomed as he was to them, not, it may be, so deftly as his opponent. Sandow, on the other hand, set three and performed two of his own difficult feats—*not one of which Hercules attempted!* How, in face of this result, the honour and rewards of the victory could go to Hercules is, the reader will no doubt say, incredible. It is only paltering with the public to take exception to the manner in which Sandow performed his work, spent as he was by the prolonged and severe trial. The manner of doing the feats was not conditioned in the articles, and, if it had been, the use of the body's leverage in elevating the weights from the shoulder is certainly more allowable—because scientific and hygienically safe—than the tricky and vicious use of the jerk. The matter, however, is in a nut-shell: Hercules did *three*, and Sandow *four*, of the six feats in the contest; while the latter essayed to do all, but, to settle another wager, was handicapped by having to perform an arduous feat prior to undertaking the competition proper. If the contest was to decide a matter of strength, which was the stronger man will be seen by a reference to the number of the tests, and more particularly, to the far heavier weights which Sandow was able to manipulate. Sandow successfully lifted in all a total of 1,007 lbs., and failed to lift another 430 lbs. Hercules, all told, only lifted 552 lbs.

XI.

SANDOW BREAKS ALL RECORDS.

THE great athlete was now to win a trophy by such a display of weight-lifting as should set forever at rest not only any question regarding the match with Hercules, but effectively put in the shade all previous records of Mr. Sandow's prowess. For weeks after the match, denunciation of the judges' decision had been raging in the Press, and great efforts were made, in which Mr. Sandow joined, to re-try the test of strength between the contestants, but without practical response from Hercules McCann or his backers. At this juncture, the London Athletic Institute, with Professor Atkinson, F.R.C.V.S., at its head, stepped generously forward and offered for competition a gold championship belt, to be awarded to the man who would make the best English record in weight-lifting. An invitation was extended to

Sandow, who, as the virtual champion of heavy-weight lifters, and known to have engaged to break all previous records, was likely to be unapproached in the coming exhibition. The following account of the evening's performance, taken from *The Sporting Life*, Jan. 29th, 1891, will show how well-nigh unsurpassable was Sandow's feats on the occasion. The exhibition took place at the International Hall, Café Monico, Piccadilly Circus, on the night of the 28th of January before a crowded and enthusiastic audience. The English record of weight-lifting to be beaten on the present occasion was that of Hercules McCann, the opponent of Sandow, in the contest in which, though not to the satisfaction of the public, as we have seen, the judges' verdict went in favour of that athlete. What the audience now assembled were to see was McCann not only beaten at his own game, and in the feats he specially affected, but the establishing of a record for Sandow which eclipsed all existing records and won for him the great prize of the evening.

GREAT RIGHT AND LEFT HAND WORK.

We take on the report of the *Sporting Life*, after introducing the subject.

"By this time Sandow was ready, and soon Herr Condol, his manager, was busy getting his heavy weights together, while masters of ceremonies, Mr. Bush and Frank Hinde, saw to the outside preliminaries. The judges consisted of Colonel Fox, Colonel Burchard, Messrs. F. A. Bettison, John W. Fleming, and J. Couttes, with Professor Atkinson, as referee. The latter also acted as spokesman, and in a few well-chosen words told how the gold belt was to be won. He said that the feats set by Henry ('Hercules') McCann at the Royal Music Hall two months ago would be considered the standard. Sandow then doffed his ulster and stood revealed in salmon-coloured tights, with a black leotard, and black leather sandals adorned his feet. While he was wiping his hands, preparatory to the warming up exercise, Shirley Jevons, one of the judges of the Sandow-Hercules contest, approached the stage and asked a question. He was instantly invited to an exalted position. Sandow, in the meantime, was toying with a pair of 100 lb. bells, one in each hand. He curled them up to his shoulder, and then held them aloft without the slightest semblance of jerk, push, or a press. The right-hand bell he elevated three times in succession, just to get his muscles wound up.

"BREAKING HERCULES'S RECORD.

"The real business of the evening was begun by the lifting of a dumb-bell weighing 179 lbs. with the right hand. Sandow stood over the mass of iron, and then getting a good grip of the handle, lifted it shoulder high. He tried to push it upward, but after geting the bell started he had to drop it to the shoulder. The second attempt was successful and Hercules's record of 170 lbs. was swept among the 'has beens,' the record being raised 9 lbs.

"Next in order came a two-handed feat. This time Sandow lifted a bell weighing 126 lbs. with his right hand and 119 lbs. with his left hand. It will be remembered that when Hercules put up his two bells of 120 lbs. and 112 lbs. he used a mighty jerk, and Sandow failed to get the bells up at all. There was no doubt about last night's attempt. Sandow got the two bells to his shoulder in very neat style. Then he started to press them up, but hesitated momentarily. The pause looked ominous, but slowly and surely the arms began to straighten and in a few seconds the two masses of iron were held aloft, Sandow not only wiping Hercules's record off the slate, but making the new one in magnificent style. Mr. Jevons seemed to be in doubt about the arm being perfectly straight, but Prof. Atkinson stated that, with such enormous biceps, it was simply impossible to get the arm like a ram-rod.

" The next task was the lifting of 160 lbs. with the left hand. Hercules got up 155 lbs., and Sandow, not knowing the knack when he met McCann, could not exert his full strength. Last night, however, he had no trouble. He first curled the weight up to the shoulder and then slowly pressed it until it was well over his head. Sandow dropped the weight, looking defiantly at those who were adversely comparing his style with that of Hercules. The glance was so disdainful and Spartan-like, that the whole house burst into a volley of applause. Professor Atkinson advanced to the footlights and said that the judges were perfectly satisfied with Sandow's performance, and that he had not only surpassed McCann's record, but won the championship belt. The trophy is a beautiful one. It is made of blue satin, heavily studded with gold plate, with medallions for names and portraits. In the centre is a massive shield, setting forth how the championship was won.

"SANDOW MAKES SOME WORLD RECORDS.

"Not satisfied with showing his superiority over his late rival, Sandow set about making some new world records. His first performance was with a long-handled dumb-bell, weighing 250 lbs. This was stood in front of him to give the performer a firmer grip, but previous to lifting the weight, Sandow asked Professor Atkinson if the stage was all right. He said 'yes.'

"'Over 400 lbs. in one spot is a big weight,' observed Sandow. He referred to the bell and his own weight. Steadying himself, Sandow lifted the bell on to his chest, and then pushed it half-way up, straightening his arms as the bell rose. He stood with the enormous mass fully extended. Dropping the bell shoulder high, he again pushed it up, and tried the performance again, but the bell turned in his hand when it was half-way up, and he dropped it to the floor with a crash that made every one's teeth jar. Next the bell was stood endwise, and with two hands Sandow lifted it to his shoulder, steadying it for a moment, and then gradually pressing the bell up, he achieved one of the grandest pieces of dumb-bell lifting ever seen. This performance not only eclipses Staar's Vienna record, but establishes what had hitherto been a doubtful performance.

"The next thing done was the elevating of a bar-bell weighing 177 lbs. Sandow had no trouble in curling this weight up to his shoulder, or in pressing it aloft. The work was so cleanly done that the spectators gave the performer round after round of applause. An ordinary plate bell of 161 lbs. was the next weight handled, and this time the left hand was used. The curling process was used to bring the bell to the shoulder, and then the press was put into operation. As the iron rose in the air a faint 'Oh!' was heard, and Sandow looked daggers at the place from which the sound emanated. It seemed to unnerve him for a moment, but getting a good grip of the bell, he held it aloft as though it were a walking-stick. When it is stated that this is 6 lbs. more than Hercules put up, the magnitude of the feat can be realized, especially as McCann had not done half the work that Sandow had gone through. These are three records that will stand for some time.

"MAKING THREE RECORDS FOR THE JUDGES.

"The officials were so carried away, that they importuned Sandow to do some special feats, and the good-natured German readily complied. He stood beside the scales, watching the weighing process, and when they omitted to weigh the two nuts that are used as fastenings on the bells, he called their attention to the oversight, remarking 'I want to get credit for all I do.' These nuts weigh over a pound each, so that they make quite a little difference in the avoirdupois. When everything was in readiness, the plate bell, weighing 70¼ lbs. was placed in position, and Sandow raised it to his shoulder. Then gradually dropping the weight until his arm was at right angles with his body, accomplished one of the greatest feats of genuine strength ever known in this or any other country. The performance will now form a world's record in the absence of any known performance of its kind. The left hand was treated to a 56 lb. lift. It was a very clean one.

"WINS THE CHAMPIONSHIP BELT.

"After a little rest, Sandow came forward for the last and probably the greatest feat of all. It was the simultaneous elevating of a 70¼ lb. weight in his right hand, and a 56 lb. weight in the left hand. Raising the pair of bells to his shoulder, Sandow held them there until every one could see that there was no trickery about the feat. Then he gradually lowered his arms to a horizontal position and held the weights out. The ring of the 56 lb. weight was down, so that no assistance could be gained from the wrist. The ease and coolness of the performance electrified every one, and for some minutes no one seemed to realize the magnitude of the achievements. When one individual did start the applause, it soon swelled in volume, and for some minutes the noise was deafening. When quietness was restored, Professor Atkinson stepped forward and presented Sandow with the championship belt, saying, 'You have not only eclipsed all Hercules's performances, but you have set a lot of tasks that will remain on record for a long time. In addition to this, you have given us an exhibition of pure strength that seems phenomenal. I have great pleasure in presenting to you the championship-belt, which I hope you will find pleasant to look at in after life, and I also hope that you may live many years to enjoy it.' Sandow's eyes sparkled as he took the valuable trophy, and he looked as if he would like to say something, but his non-familiarity with the idioms of our language kept him silent, and he could return thanks only with his frank blue eyes."

XII.

PHYSICAL CULTURE IN ITS RELATION TO THE ARMY.

THE presence and successes of Mr. Sandow in England naturally quickened public interest in all manner of gymnastic exercises, and directed afresh the attention of the military authorities to physical culture, on the great athlete's system of training, in its bearing on recruits for the army. Sandhurst, Woolwich, and Aldershot, all felt the influence of the vogue for muscular development aroused by the exhibition of strongmen in the metropolis. One of the most enthusiastic of Sandow's admirers is Colonel Fox, Inspector of Military Gymnasia for the British Army and Director of Physical Training at Aldershot. This officer had become much impressed with Sandow's phenomenal muscular proportions and enamoured of his system of training, which produced such results as the renowned athlete exhibits in his person. Examining critically into the system, Colonel Fox assured himself of its simple yet effective methods, and in repeated interviews with Sandow obtained from him such hints as has induced the gallant Colonel to adopt his exercises in the training schools for the army. The recruit of the future, Colonel Fox determines, shall be a man ready trained for campaign-work, not, as has too often happened in the past, a man whom the campaign has to train.

Imbued with these views, Colonel Fox took advantage of such occasion as presented itself to bring Sandow as a model before instructors and cadets in the military training schools; and in this good work he was fortunate in enlisting the co-operation of not a few of the medical staff in the various depots of the army. One of the most intelligent and devoted among the latter is Surgeon-Major Deane, of the Medical Staff, who, on the 12th of December last (1892), delivered a lecture on Physical Culture at the Royal Military Academy, Woolwich, taking advantage of Mr. Sandow's presence to give point to his lecture in illustrating what he had to say on the subject of gymnastic anatomy. The lecture was so important, and interesting from the fact we have stated, that an account of it was published in the London *Lancet* (Dec. 24, 1892)—the chief organ of the medical profession. We transcribe the report verbatim, deeming it of much interest to the intelligent reader:—

"AN OBJECT LESSON IN GYMNASTIC ANATOMY

"On Monday, the 12th inst., a lecture on Physical Education was delivered in the Gymnasium of the Royal Military Academy, Woolwich, by Surgeon-Major Deane, of the Medical Staff. The lecture, which had been previously given at the Royal Military College, Sandhurst, was in itself well worth listening to, but it excited a good deal of popular interest—as far as the cadets were more especially concerned at any rate—owing to the fact that Sandow, the strongman, was in attendance and afforded in his person a practical illustration of what can be done by physical training in an individual naturally of powerful build—in fact, an object lesson in gymnastic anatomy. The proceedings were under the auspices of Colonel Fox, the Inspector of Gymnasia at Aldershot, and there was, it need scarcely be added, a full attendance. The lecturer commenced by giving various instances in ancient, mediæval, and modern times of men who were characterized by their superior development of both physical and mental qualities, ending by citing the present Prime Minister, 'as not only a man of powerful intellect, but as a hewer of trees.' He then went on to explain that nature had given us a certain amount of capital or reserve on which we could draw, and added that this might be more clearly represented by assuming that our personal equation was 1. This reserve force was continually being drawn upon, and could only be maintained by good food, sleep and healthy exercise both of mind and body. He pointed out that physical exertion and exercises undertaken for strengthening and developing the muscles were not without exercising a favourable influence also in developing the mind, and among other illustrations remarked that it was commonly recognized that the more exercise a schoolboy took, the more fresh and quick he became in his studies. Be this as it may, however,—and in a sense and within limits it is undoubtedly true,—the lecturer proceeded to say that if England was the most athletic nation it was also the worst physically trained one, for young men took up such games as cricket, football, racquets, or running, which collectively were very good indeed in their way, but he pointed out that, taking them separately, they all tended to develop only certain parts of the body. In order to avoid this partial development the first thing to be noticed in studying the human frame is, that it is made by nature to stand erect, from which we might infer that all exercises should be performed in that position on the ground on which we stood, and not above it, as in so many of the exercises provided in gymnasia in England. Sandow's development had been attained by constant and systematic use of the muscles, and especially by the employment of 5 lb. dumb-bells, each exercise being designed to increase the power of some particular muscle or group of muscles. Sandow had modelled his system of training on that in fashion with the Greeks and Romans. He had not employed any modern gymnastic apparatus, but had attained his marvellous muscular development mainly by the use of light dumb-bells in connection with observations on the anatomical arrangement and disposition of his muscles.

"The lecturer then asked Sandow to perform certain feats and exercises in illustration of what had been advanced. From this point to the conclusion, the proceedings became, in a physiological and anatomical sense, very interesting and instructive, for rarely indeed can the various muscles be seen by being put into action in the living body as definitely and precisely as if they had been laid bare by a dissection in a dead one, as was the case in Sandow's exhibition of them. Stripped to the waist, he was able to demonstrate by different movements how great was the command he had over various muscles. Clasping his hands behind his head, he was able to make his biceps rise and fall in time to music. Walking round the audience, he displayed various muscles in action as they were separately named. By putting his hand behind his back in such a position as to cause contraction of the deltoid, he can raise that muscle to a degree that makes the shoulder look out of all proportion to the rest of his body. The development of the flexor and extensor muscles of the upper extremities, especially of the triceps, was also noteworthy. He can flex or bend his wrist to such an extent that a vertical line drawn from the knuckles will fall on the region of the muscles of the forearm. The intimate physiological connection between the terminal nerves distributed on the skin and those of the muscles beneath, as well as the contractile power of the muscles themselves, are readily manifested; and the normal reflexes should be capable of being easily demonstrated. Sandow applied the hands of some of the bystanders to the skin over the chest walls and other parts of the trunk of his body, with the result that a young fellow described the sensation as being like that of 'moving your hand over corrugated iron.' Standing in the centre of the room he showed his maximum and minimum chest measurement. After an efforted expiratory act, aided apparently by the pressure of his arms against the ribs laterally, a difference of twelve inches is caused by deep inspiration and forcible action of the inspiratory muscles. When he fully inflates his chest and 'sets' its muscles, his arms form an angle of about 40° with his body, owing to the size and prominence of the muscles under the arm and towards the back of the shoulder and those of the lateral aspect of the chest. The pectoral and serrati muscles are very noticeable. Taking two packs of cards together he attempted to tear the two packs—104 cards—in twain, and, after spending about ten minutes in his efforts to do so, he succeeded in accomplishing his purpose, affording at the same time an indication of the great muscular strength of the hand and wrist. He failed in doing this at Sandhurst. In order to illustrate the development of the muscles of the back he took a short length of circular india-rubber of about an inch or more in diameter and fitted with handles. This, on being previously passed round the audience, could hardly be stretched by four cadets pulling at each end. Sandow, however, taking hold of the handles and turning his back to the audience, stretched the india-rubber across the back of his neck until his arms were extended at right angles to his body. The action of the muscles of the back caused them to look, as it was remarked, like snakes coiling and uncoiling themselves under his skin. In order to show his weight-lifting power he used a bar-bell weighing 270 lbs., which one of the strongest sergeants of the academy had only succeeded in lifting from the ground by the use of his body as well as his arms. Taking the bar-bell in the centre, Sandow allowed it to swing, as it were, by its own weight across his shoulder, from which position he slowly raised it upwards to arm's length above the shoulder. An arrangement was then shown for exercising the adductor muscles of the leg. It consisted of two upright posts and pieces of india-rubber, which are hooked to them and to straps which fasten round the leg just above the knee. The performer sits in a chair between the posts and tries to press the knees together by extending the india-rubber. A cadet who had tried the apparatus could with great effort just do this with three pieces of india-rubber connecting his legs with the posts. Sandow, having attached one more piece of india-rubber on each side, which was all that was available, opened and closed his knees with the utmost ease and without any apparent effort. With the view of showing his gymnastic agility, Sandow very neatly turned a somersault at the close of the performance. His personal equation, as compared with that taken on the previous assumption, may be represented as 50. It is scarcely necessary to add that, with cadets for an audience, Sandow did not lack applause and that there is at present a 'great run' on all the light dumbbells at the Royal Military Academy. The demonstration is, as we have said already, chiefly interesting from an anatomical and physiological point of view, and we have not attempted to discuss the merits of his system from the standpoint of military training and hygiene. The advantages of out-door exercises and sports—in the way of fresh air, emulation, pleasurable excitement and variety—over more systematic and exact methods of physical training need not be stated, for they are obviously on the side of the former."

The interest manifested at Woolwich in Sandow's personality, and in his effective system of physical education, was also manifested at Aldershot and other regimental depots and places of military training throughout the British islands. Mention has already been made of the fact that army men generally had viewed with lively enthusiasm Sandow's exhibitions of feats of strength, and that his methods of physical instruction had been adopted by the military authorities. One of the most alert and intelligent of British officers to confer with the great athlete on his system of training was Lt.-Col. G. M. Fox, Her Majesty's Inspector of Military Gymnasia for Great Britain. This gentleman made Sandow's acquaintance shortly after the latter came to London to begin his successful professional career, and from the first was interested in his methods of physical training and impressed by his redoubtable achievements on the stage. Colonel Fox's own efforts had been long and earnest in seeking to improve the physique of recruits for the army, and his labours in this direction have, admittedly, borne much good fruit. Naturally, the gallant colonel took an interest in Sandow's advent in London, and he made it his business, as we have already said at the opening of this chapter, to inquire closely into the system of exercises by means of which the strongman had made himself strong. Learning what these exercises were, and the success which attended the observance of the simple rules which Sandow imposed upon himself in training, Colonel Fox put both to practical test, with gratifying results in the sphere of his important duties. In obtaining these effective and pleasing results, Colonel Fox was aided by Sandow's presence at Aldershot, and by his "coaching" of the Staff Instructors and non-commissioned officers under training at the depot.

While this volume was under way, Colonel Fox was written to by Mr. Sandow requesting such information as he, in his official capacity, might deem it proper to give, anent the success which had attended Mr. Sandow's training instructions, and that officer, with ready and friendly courtesy, instantly complied with the request. The reception of Colonel Fox's letter was naturally gratifying to the great athlete, and especially so as the testimony comes from an able and distinguished British officer, known for his zealous efforts in helping to raise the standard of physical efficiency in the army. The letter, which is subjoined, we have the kind permission of its writer to publish. Here it is :—

AN ENGLISH LIEUT.-COLONEL ON SANDOW'S METHODS OF TRAINING.

"THE GYMNASIUM, ALDERSHOT,
"29th July, 1893.

"DEAR MR. SANDOW,
"I am in receipt of your letter from New York, which

reached me on the 23d inst. and am very glad to hear of your success in America. The book you speak of as being about to be published, should also be very successful, and ought to do much towards making your system of physical development widely-known. Since your last visit to us here my Staff Instructors and non-commissioned officers under training have been most energetically practising the light dumbbell exercises you so kindly showed them.

"I am convinced that your series of exercises are excellent and most carefully thought out, with a comprehensive view to the development of the body as a *whole*. Any man honestly following out your clear and simple instructions could not fail to enormously and rapidly improve his physique. As two notable instances, I may cite the cases of Captain Woodgate, Superintendent of Gymnasia, Woolwich, and of Staff-Instructor Moss, Army Gymnastic Staff.

"It is almost superfluous for me to add, that you yourself, *in propria persona*, are the best possible advertisement of the merits of your system of training and developing the human body. Perhaps the best part of your system that *I* think most highly of, is your insistence (1) upon the concentration of the will-power on the muscles or muscle chiefly concerned in an exercise; and (2), the importance you attach to energy and dash, accompanied by the most rigid attention to the minutest details, in the actual carrying out of any and every exercise. As the result of twenty-five years experience, I can confidently assert that work done without strict attention to these two points is valueless, from either a developmental or educational point of view, if, indeed, it be possible to differentiate between terms that are, *a priori*, of necessity almost synonymous. It is of course extremely difficult, and well-nigh impossible, to insure the concentration of will-power upon an exercise among large masses of men, whose physical training is more or less compulsory; and we have then to fall back upon the expediency of fixed apparatus to insure the attainment of the necessary amount of muscular exertion. But any individual, gifted with a fair amount of determination, is absolutely certain to develop his physical powers at an extraordinarily rapid rate and with the most happy results to his general health and mental powers and activity, by following with intelligence your system. As you very rightly say, it is only by bringing the brain to bear upon our exercises that we can hope to produce the best results with the shortest possible expenditure of time.

"The absence of expensive and cumbrous apparatus is no small recommendation of your system, and you are thoroughly in the right when you assert that lasting muscular development, and consequent strength, can be best produced by the constant and energetic use of light dumb-bells employed in a sound and scientific manner.

"Believe me, yours very truly,
(s) "G. M. Fox, *Lt.-Colonel*,
H. M. Inspector of Military Gymnasia in Great Britain.
"Professor Eugene Sandow,
"New York, U. S. A."

This, the testimony of a high and competent authority, to the importance of Mr. Sandow's methods of physical training will, no doubt, be received at its proper value, supported as it is by the practical tests to which the system has been put. From other military sources, and especially from many zealous regimental instructors, Mr. Sandow has also received equally emphatic endorsement of his intelligent and effective system. Its fruit is, moreover, shown in the announcement, recently made, that the Commander-in-chief of the British army has sanctioned the introduction of light dumb-bells and kindred appliances of athletic training, and approved their use, in the various gymnasia at the home-depots of regimental districts and cavalry riding schools.

XIII.

SANDOW "AT HOME" AND ABROAD.

The title of this chapter is chiefly to record an incident, of an amusing kind, which happened to Mr. Sandow as one of the many millions of lodgers in the great metropolis. What we have to relate is a case of "bringing down the house"—though not quite in the professional sense—and the consequent proof that the law, in England at any rate, is even stronger than the strongman. We shall not spoil the story by drawing upon the bald recital of what happened, from the police court records, but leave the reader to gather the facts from a sprightly editorial in the London *Globe* of Nov. 13th, 1890. All we need say, is that Mr. Sandow, while one day exercising in his rooms, did direful damage to the ceilings, walls and furniture in the house in which he abode, and not agreeing quickly with his landlady while he was in the way with her, was summoned before the judge to atone for the wreck he had occasioned and be admonished to rehearse his feats, for the future, in some lonely, sequestered spot. Says the *Globe* :—

"SANDOW CHEZ LUI.

"The strongman has been very much in evidence of late years, but little light has been hitherto let in upon him in his domestic relations. Yesterday's law reports, however, go some way towards supplying the deficiency. The case of Brackenbury *v.* Sandow, decided in the Westminster county-court the other day, will be perused with interest by all lodging-house keepers. This class of the community are commonly supposed to be able to take excellent care of themselves, but from the present case it is clear that even a London lodging-house keeper is capable of entertaining a strongman unawares. We say this advisedly, for it is obvious that no landlord, unless his house was built specially for the purpose, would be so rash as to welcome in the capacity of a lodger a gentleman who is in the habit of sporting with dumb-bells weighing 312 lbs. Having done so, however, and having been so indiscreet as to house the strongman on an upper storey, the landlord in question soon realized in a very practical way the risks to which he had exposed himself. The ceilings and his patience gave out about simultaneously, and litigation set in, with the result that Mr. Sandow, who did not appear, was ordered to pay damages to the extent of £4, 12*s.* 6*d.* Hitherto the professional musician has been the chief *bête noir* of the landlord, but now the strongman must be added to his *index expurgatorius.* The moral of the episode is fairly transparent. Always ascertain, if your calling be that of a letter of lodgings, whether your intended lodger be a professional follower of Hercules or not, and in the event of his being so, never offer him quarters except in the basement. Perhaps in the 'ideal flats for professional men,' of which we have heard a good deal of late, suitable provision will be made for tenants of this description."

Sandow's possession of the crown of strength was, about this time, amusingly—perhaps even tragically—illustrated in another way. He had run over to Paris on a short holiday, and there met an old schoolmate staying at the Grand Hotel. After a long chat over old times and the memories of their boyhood days, the friend suggested a game at billiards, which Sandow agreed to, adding, however, that he was quite out of practice and would be found but a poor player. This, in his friend's eyes, was of little moment, for, as he said, the pleasure of being together again would give sufficient interest to the game. The billiard-room was crowded and it was with difficulty the two old schoolfellows obtained a table. They hadn't been playing long when a party of Frenchmen came and stood alongside, evidently eager to get possession of the table. One of the number, observing Sandow's indifferent playing, made a rather offensive audible remark, which Sandow's friend resented, but Sandow himself interposed and prevented the altercation going further. Later in the evening, the two friends retired to the restaurant for supper, and when they had taken their seats they found themselves in close proximity to the party of Frenchmen with whom they had all but come into collision an hour or two before.

SANDOW CHASTISES A BELLICOSE FRENCHMAN.

During supper, when the wine began to flow, one or two of the Frenchmen became first hilarious, then daring and saucy. Sandow and his friend had taken little notice of the party until a remark was made by one of the French roysterers, pointed at the young Germans, and conveying an insulting reference to their alien tongue. At this, Sandow's friend, becoming angry, shot a retort back at the Frenchmen, when one of the latter jumped up and menacingly shook his fist at the Germans. Sandow motioned the excited Gaul to sit down, telling him, in French, that it would be better for him to keep quiet. There was something in the nonchalant way in which Sandow had given this counsel that irritated the Frenchman, and he crossed to the Germans' table and gave Sandow a blow in the face. His friend squared up at this outrage, but Sandow again interposed and coolly turned to the Frenchman and cautioned him, at his peril, not to strike again. He did strike, however, and, this time, with a sharp blow on Sandow's nose, which set it bleeding and stained a new light suit of tweed which the athlete wore. So far, Sandow had put a rigid restraint upon himself, but angered at the soiling of his clothes, and to keep his friend from engaging in a general tussle, he, in an instant caught the Frenchman by his legs and the back of his neck and brought his knees into repeated and ignominious contact with his nose. He then rapped his fundament on a table with such force as to break the latter and set his foolish aggressor unconscious on the floor. The chastisement was the work of a minute, but it sufficed the now alarmed Frenchmen, who were dumfounded at the sharp and unexpected reprisals and felt that their friend's attack was unjustifiable and unwarranted. Their concern, however, was great for their prostrate companion, who had to be taken to an hospital, while Sandow and his friend gave themselves up to the gendarmes whom the waiter and his master had summoned.

For two weeks after his admonishment by the angry athlete, the titled Frenchman—for it transpired that he was of high birth—languished in an hospital ward, inwardly profiting, meanwhile, by the lesson that had been administered him. Sandow's explanation to the police saved him from imprisonment, and, regretting the severity of the chastisement he had inflicted, he did not fail, while he remained in the gay capital, to call daily upon the now penitent, but not convalescent, aggressor. The incident had a sequel, which we have now to relate.

SANDOW RECEIVES THE GIFT OF A GOLD CHRONOMETER.

One evening, while exhibiting at the Tivoli Theatre, on his return to London, a card was brought to Sandow from a gentleman seated in one of the boxes accompanied by a party of friends. On the card was penciled 'the admiring homage' of the gentleman whose name it bore, with the request that Mr. Sandow would honour the party with his presence at the close of his performance. Mr. Sandow complied and was warmly received by the gentleman and his friends, who extorted from him a promise that, after his bath at the close of the exhibition, he would join the party at supper at the Hotel Savoy, whither they proposed to adjourn. There he learned that the gentleman who had pressed upon him the invitation was he with whom he had had the encounter at Paris! This gentleman, who had only through his visit to the Tivoli discovered his erstwhile chastiser, was now profuse in his apologies to him for his previous rudeness; and with the utmost frankness and cordiality he explained to his friends the motive he now had to make atonement. Mr. Sandow met his host in the same spirit of amity and greatly enjoyed the evening he spent with him and his friends. Next day, at his rooms the strongman received by the hand of a valet a little box, which on opening he found to contain, besides a polite note begging his acceptance of the souvenir, a gold chronometer, by Bennett, of very considerable value, with a combination of ingenious mechanical adaptations, for striking the hours, minutes and seconds, a perpetual calendar, and other curious and elaborate contrivances. The gratification of Mr. Sandow may be imagined, for the handsome gift, it need hardly be said, came from his Parisian friend, whom he had once used so roughly. The chronometer, we may add, is the great athlete's daily companion and one of the most highly-prized of his souvenir possessions.

Another incident, of an amusing kind, may here be cited to illustrate how ugly a customer Sandow may be found should occasion call for the exercise by him of his strength. The "noblest Roman of them all"—if the phrase will be pardoned—had been spending a holiday, in the spring of 1892, in some of the cities of Italy and Southern France. If the truth must be told, he had been beguiled to Monaco, where he had won at that gambling resort 25,000 francs, and, as a matter of course, had also speedily lost that sum with considerable additions to it. As he was wending his way back to England, he had occasion to stop at Nice, where he had had considerable personal effects, consisting of about two thousand pounds' worth of jewellery including prize medals, souvenirs, and other valuables, which he desired to have sent on to London. The whole were packed in a trunk and sent to the railway station at Nice for its despatch to England. Sandow had himself come to the station to arrange for the transmission of the box; but before conferring with the agent he was accosted by two men on the platform who proffered their services as interpreters, and so led the railway people to infer that they were friends of the athlete. Sandow, however, did not require their services, as he himself spoke French, and he turned from them to the porter and gave his own instructions for the despatch of the trunk, getting into a carriage as he did so, and left for Paris. In due course, he arrived at the gay capital, and there made a halt on his journey. While there he learned from his agent at London that the box had been received, but, on opening it, it was found that the valuables had been abstracted, and their weight partly substituted by half-a-hundred of bricks! On receipt of this startling intelligence, Sandow at once returned to Nice and instantly sought the railway porter to whom he had intrusted his valuables. From this person he learned that when he had set off for Paris, the two men who had addressed Sandow on the platform, and whom the porter had taken for his friends, had come to him, as they said at Mr. Sandow's request, and got possession of the box, saying that they had his instructions to forward it through another channel. The porter, not doubting the story, delivered the box, and the men drove off with it—the last the railway people had seen of it. Provoked at the way they had been imposed upon, the railway authorities placed the porter at Mr. Sandow's disposal in the efforts now made to get on the track of the depredators, who were supposed to be still in the neighbourhood, and endeavour to recover the lost possessions. This assistance, after a day or two's search, was effectual, and the thieves were espied on the street. Sandow, who, meanwhile, had refrained from calling the police to his assistance, now acted without their aid. He pounced upon them suddenly, and caught each man firmly by the back of the neck. When they recovered from their surprise and began to struggle to get free, the strongman brought their two heads repeatedly in contact, until unconsciousness rendered one man limp and fright quieted the other. Without quitting his hold of the men, Sandow dragged them both to the station, into which he flung them, to the surprise, and amusement of the police. It took some days for the miscreants to recover their senses and appear before the court: in the meantime, they owned to the crime they had committed, and on their persons were found the pawn-tickets which enabled Sandow to recover his impounded effects. With the recovery of his property he refrained from prosecuting its despoilers, content that by his rough handling of them, the reader will say, he had taught them a sharp enough lesson. It is something to be one's own law-enforcer.

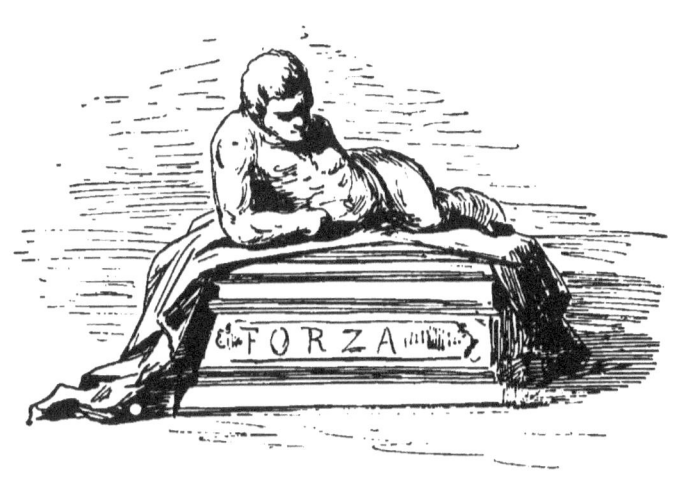

XIV.

SANDOW IN THE NEW WORLD.

THAT Mr. Sandow, a man of such mighty muscle, with unparalleled drawing powers, should be tempted of *impresarios* to fill a golden engagement in the New World, will be taken as a matter of course by readers of this book. Sandow's nationality was in itself a drawing card, for the German element is large in the United States; large also is the class within the Republic that takes a lively interest in athletics. These several facts were doubtless known to Mr. Henry S. Abbey, who made the contract with Mr. Sandow for a lengthened engagement on this side the Atlantic. Hence it was with no surprise that we heard of the renowned athlete's departure from England to make a professional tour of this Continent. Nor were we surprised on other grounds, for the young Prussian, incited by youthful ambition, and possessing the energy and enthusiasm of his nation, was himself desirous of seeing the New World and its people ; and so he readily embraced the overture made to him by the well-known and enterprising theatrical manager. The result of the agreement to both interested parties has already justified the anticipations each looked for from the visit ; while public interest, whetted as it had been by the Old World fame of the great athlete, has, so far, in the three chief cities of the United States, been widely gratified.

Mr. Sandow opened his American engagements at the Casino, New York, in June of the present year (1893). He has subsequently appeared at the Tremont Theatre, Boston, and at the Trocadero, Chicago. From the first his exhibitions have been entirely successful, despite the fact that he arrived at the close of the theatrical season. Drawn to them were not only large and delighted audiences, including thousands of sporting men and amateurs and professionals devoted to the study of athletics, but crowds also of medical men, physiologists and anatomists of note, who viewed with critical but admiring eye the great athlete's wonderful muscular development and surpassingly fine physique. Instructors and pupils from the New York gymnasia and from all manner of athletic associations came to the Casino in full force and were enthusiastic in their applause of Sandow's varied feats. Nor was the Press, with its wonted enterprise and ready intelligence, less cordial in its reception of the wonderfully-endowed newcomer, whose advent was hailed with such general and hearty acclaim. Notable among the journalistic greetings of Sandow in the New York Press were those which appeared in the great metropolitan organs, *The World* and *The Herald*. Each of these newspapers devoted much space, in successive issues, to biographical and professional facts respecting the renowned strongman, with accounts of interviews and other descriptive matter bearing on Mr. Sandow's rare physical endowment and extraordinary performances. *The World* published two such articles, both illustrated, one chronicling Sandow's feats and giving an abstract of his various Old World achievements; the other detailing a physical examination of the strongman by a scientific expert, Dr. D. A. Sargent, Director of the Hemenway Gymnasium, Harvard University. *The Herald* also published an interesting illustrated article on the *début* of "this modern marvel of physical power, beside whom the average man is puny,"—to quote the journal's apt characterization,—with the recital of an interview by its reporter. From these representative newspapers we shall take the liberty of drawing some facts of probable interest to the reader.

THE NEW YORK WORLD ON SANDOW.

In the first of *The World's* articles (June 18, 1893), its readers are thus introduced to Sandow. We quote from a passage in which the reporter has expressed the opinion that while nature had set out to make "a conspicuously fine job" of Sandow's physical frame, he had, by training, "made himself a great deal better man than Nature intended him to be." " In preparing the mind for a description and conception of this wonderful human being," says *The World*, " it is necessary to abandon all former notions concerning possibilities in physical development. Nothing that has ever been seen in New York can be used as a standard of comparison to measure the wonderful young German who has just come here. Compared with Sandow, Corbett, the fighter, is like a lean spring-chicken beside a well-muscled bull-dog, and the professional strong man of circuses and museums, with their pretentious bunches of muscle, seem weak and unimpressive.

"A proper way to introduce Sandow is to outline briefly some of the things which he can do. Sandow can lift a 500-pound weight with his middle finger. He promptly took up in London an individual who bet that he could not perform this feat.

" He can break good-sized iron rods across his arms and legs, but does that rarely because he considers the achievement trivial. He takes in his right hand a dumb-bell with an enormous sphere at either end. In each of these spheres a man is concealed. He lifts the dumb-bell and the two men above his head with one hand.

"He can take a good-sized man with one hand, and without any sign of effort use the man's body for a musket and give an imitation of a regulation drill. He can oblige any friend he has in the world by letting the friend sit on the palm of his hand and then lifting him in the air above his head as easily as the average man would lift a small-sized dog.

"He places himself upon the floor with his chest upward and supported only by his hands and feet, his body forming a bridge. A gang-plank is placed across his chest and three horses stand upon this at one time, with no support except that which the chest offers. Two of the horses are small horses and the third is not enormous; but the weight of the smallest horse would more than satisfy the chest of the ordinary strongman.

"He has wrestled with three men at one time, all expert wrestlers, all bigger than he, and has stretched first one and then another flat, using one hand to a man and incidentally preventing the other two from tripping or otherwise throwing him.

"Sandow's actual feats of strength, however, do not make up his strongest claim to attention and veneration. The great point is that the man who does all of these things is only 5 feet 8¼ inches high, and does them because he has developed to the highest point every separate muscle in his body.

"There are thousands of men in the world who would tower from six inches to a foot above him and who weigh nearly twice as much, but it is not likely that any one could equal even the sheer brute strength of this German bunch of muscle which weighs exactly 200 pounds, is 5 feet 8¼ inches in height, and within 2 inches of 5 feet around the chest, when fully expanded.

"The measurements of the man's chest and waist perhaps give the best conception of his wonderful conformation. Around the waist he measures twenty-nine inches; around the chest, when fully expanded, as has been said, he measures fifty-eight inches; his waist, therefore, is not much bigger around than Mrs. Langtry's, and his chest is a good deal bigger around than Grover Cleveland's. Grover Cleveland, Mrs. Langtry, and the entire public must be interested in such figures as these. They are based on accurate and careful measurements. It is needless to say, that when this young man spreads out his chest and draws in his waist, his body from the shoulders down to the hips, looks like a very sharp wedge of pink muscle. The writer, who called upon Mr. Sandow and examined carefully his mental and physical make-up, has had the pleasure of studying numerous types of the muscular human being. He has studied the finest specimens of manhood to be found in the German gymnasiums, but he experienced an entirely new and unexpected series of sensations upon beholding Eugene Sandow.

"In private life this young man is a very pleasing type of the simple-minded German. His head is shaped exactly like the heads on the old statues of Hercules. The forehead is low and rather broad. The head is not quite straight up and down behind, but with only slight development. It is thickly covered with a short crop of tight golden curls, each one looking as though it had been specially, fixed up with a hot iron; but the curliness is perfectly natural. The impressive muscular feature about Sandow, as seen fully clothed, is his neck. This neck, which is padded on either side with muscles about as big as a young girl's wrists, is nearly twenty inches round, almost as big round as the head above it. It wouldn't be a bad neck for a small bull. It is a wonderful neck for any man. His face is a pleasant face; his eye, which is gray, shows the character which has made him the man he is and which enables him to attempt with absolute confidence and calmness the various feats that fill his audiences with delight and make him rich.

* * * * * *

"Sandow has a method of his own to develop the muscles. It consists in various exercises with two dumb-bells weighing five pounds each. He declares that with these dumb-bells he has developed, not only the muscles which everybody can see on the outside of his body, but internal muscles which strengthen the walls of his chest, enable his heart and other organs to endure great strains and assure him a long life. He does not take special care of himself in the way of eating or drinking. Beer and wine are not strangers to him and tobacco is his intimate friend. He leaves brandy alone, however, as he does similar poisons. An interesting feature of Sandow's method of training is that he can train very well sitting on a chair. He can sit down and read a paper and keep his muscles working all the while, so that all development of fat is rendered impossible and his strength is kept up to the highest pitch.

"Sandow is living now (June, 1893), at No. 210 West Thirty-eighth street. With him there lives a friend, Mr. Martinus Sieveking, who is a very able pianist. Mr. Sieveking is a Dutchman. His musical compositions have already attracted considerable attention in London, and he is an unusually brilliant artist. He and Sandow are bosom friends. He thinks that Sandow is a truly original Hercules, and that no one has ever lived to be compared to him. Sandow thinks that Mr. Sieveking is the greatest pianist in the world and that he is going to be greater. It is pleasant to see them together. Mr. Sieveking, who is a very earnest musician, practices from seven to eight hours a day on a big three-legged piano. He is decidedly in earnest. He practices in very hot weather stripped to the waist. While he plays, Sandow sits beside him on a chair listening to the music and working his muscles. He is fond of the music, and Sieveking likes to see Sandow's muscles work. Both enjoy themselves and neither loses any time.

"Mr. Sandow, at the suggestion of his friend, Mr. Sieveking, was kind enough to demonstrate the fact that even in his every-day apparel it is possible for him to manifest his strength. He held up his right hand and requested the visitor to grasp his forearm. Then he closed his hand and bent his muscle till a lump rose up on his arm above the wrist which was certainly as big as a very large orange. That lump represented the force which Sandow could put into the act of closing his fingers. A feature of this young giant's life is the constant desire of those with whom he comes in contact to compete with him in some way or other. Since his arrival in

New York he has already had one challenge—which, however, was withdrawn.

* * * * * *

"Sandow's performance began on the Casino stage at 10:30. It followed the performance of Dixey. Incidentally a fine chance to compare Sandow with the average, well-developed man is offered each night. Dixey, as Adonis, at the end of his performance takes his place on a pedestal and poses as a statue. The curtain goes down and rises again to reveal Sandow also posing. New York has come to look upon Dixey as a fairly well-made young man. When New York has seen Sandow after Dixey, however, New York will realize what a wretched, scrawny creature the usual well-built young gentleman is compared with a perfect man. Sandow, posing in various statuesque attitudes, is not only inspiring because of his enormous strength, but absolutely beautiful as a work of art as well.

"One look at him is enough to make the average young man thoroughly disgusted with himself, and to make him give up his nightly habit of standing in front of his glass in his pajamas and swelling his chest with pride. Sandow's performance showed what swelling the chest can amount to when it is properly done. He expelled the air from his lungs so that the walls of his chest collapsed and his body seemed to shrink together. Then he gradually began to fill himself with air and to swell out the muscles of his chest. The development was so tremendous that it was almost painful to look at. Below his arm-pits the muscles swelled out so that his arms were forced outward and hung at an angle of 40 degrees with his body.

"The regulation performance that Sandow goes through with now is lifting two men hidden in a dumb-bell above his head with one hand, allowing three horses to stand balanced on his chest, playing with heavy weights, and lifting a man up in an extraordinary way by the muscles of his back, a feat which is called in the programme 'the Roman Column.' To prove that agility accompanies his great strength, he takes in each hand a weight of fifty-six pounds and, with his feet tied together and his eyes blindfolded, turns a somersault backward.

"Five minutes after the curtain went down Sandow, clothed only in his muscular development, was found crouching in a rubber bathtub in his dressing-room, while an attendant with a rubber pipe doused him with cold water. That was the chance to study Sandow. At first he appeared annoyed because the end of his performance found him in a perspiration. He wished it to be understood that it was not his performance of lifting two men with one hand or holding three horses on his chest that made him perspire. It was the heat on the stage, and he called up his assistant as witness. The assistant, who had nothing to do but to help half a dozen other men to carry weights, was wet through with perspiration. This fact relieved Sandow's pride. He said that in winter he never perspired at all, and that he did not strain himself.

"Taking his visitor's hand he placed it upon his heart, which had lately helped to support three horses, and called attention to the fact that there was no violent beating. In fact, the action of the heart could not be felt at all through the thick coating of muscle.

"SANDOW'S GREAT HITTING POWER.

"When he had had his bath, Sandow, with the fond pride of a mother displaying a large family of children, proceeded to display his collection of muscles, one at a time, and to dwell modestly but lovingly upon their merits. He held up his right arm and made the various muscles move about. The picture of the arm, which is often reproduced, gives but a faint conception of what it is in real life. There are very few men in New York who have as much muscle in both legs as Sandow has in that arm. The marvellous thing about it is the development of the triceps. It is the triceps which is used in extending the arm and giving a blow. The triceps in Sandow's arm is very much bigger than the calf of an ordinary strongman's leg. Sandow called attention especially to his triceps, because at the Manhattan Club he had been asked whether his great exertions had not made his muscles stiff and hard, thus rendering him incapable of hitting a hard blow. He showed tremendous speed in his movements in illustrating his hitting power, and incidentally declared

SANDOW—A STUDY.   Sarony—P

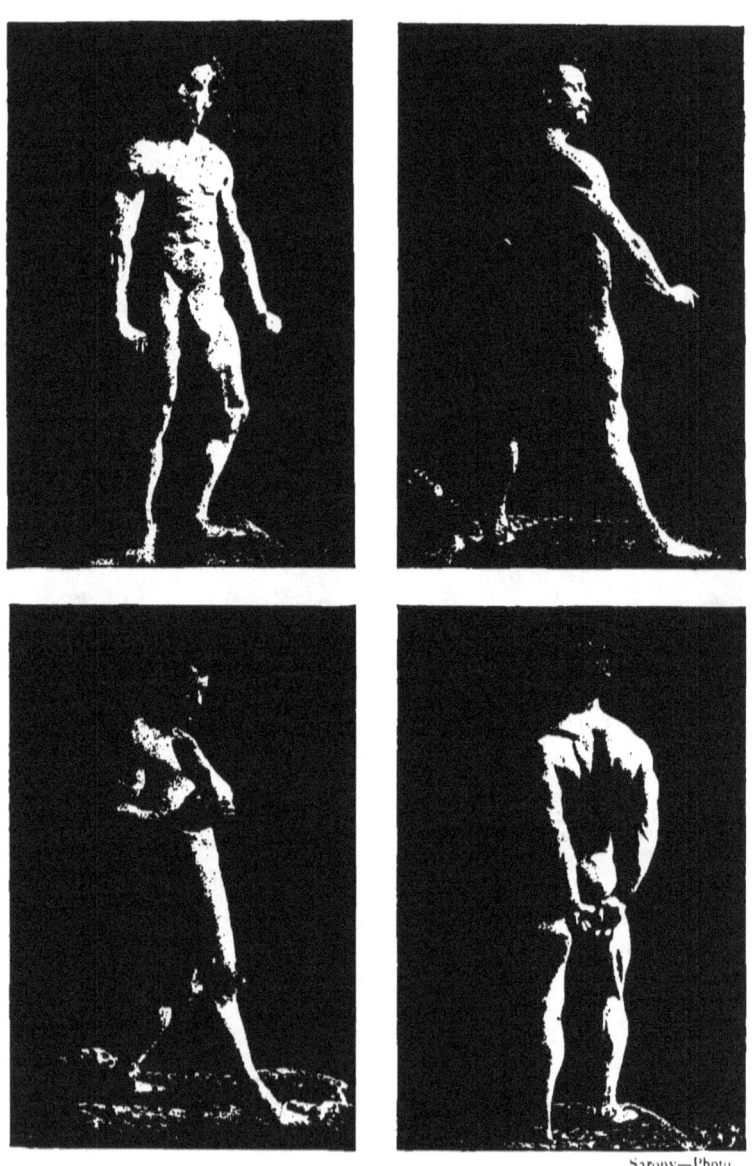

SANDOW. CLASSICAL POSES.

Sarony—Photo.

that he would undertake, with his knuckles protected, to drive his fist through a two-inch board. There is no doubt that he could do it. There is also no doubt that he could kill any man with a blow very easily. He could crush in the chest, break the neck, or fracture the skull of any man, and not use one-half his strength. Sandow was informed that in this country men got as much as $40,000 for a single fight. He admitted that that was a shorter road to wealth than the 50,000 dollars a year which he makes by exhibiting himself, but declared that he never would be a prize-fighter. 'You can't engage in a prize-fight and be a gentleman,' said Sandow. 'I care more about keeping my friends than making money.'

"Sandow went on to call attention to certain muscles which in most men are but slightly developed or have practically no existence. He swelled out his chest, and on either side of it five big muscles rose up. It looked as though five ribs on either side were coming through the skin. As a matter of fact, the ribs were not visible. What was seen was the muscle which lies over each rib, and which on the ordinary man is entirely undeveloped. Each of these muscles was twice as big round as a man's thumb, and the five on each side stood out as distinctly as though a great hand had been placed on either side of the athlete's chest.

"Next, the strongman pointed with pride to a muscle on the outside of his leg just below the waist. Each muscle, as he came to it, he called by its scientific name, for Sandow has studied medicine at Brussels, and understands anatomy. This particular muscle on the hip with most men amounts to nothing. In Sandow's case it is about as big as the leg of an old-fashioned rocking-chair.

"To show the muscles of his back, Sandow stood erect with his arms behind his head. The way the muscles are piled up on his back is most ingenious. They are so thick, so deep, that the backbone, which is quite invisible, runs along at the bottom of a deep gorge, which extends from the nape of the neck to the loins.

"Mr. Sandow was especially pleased with the muscle which he called his trapesius muscle, that is the muscle which runs from the neck over the shoulder to the top of the arm, and which accounts for the fact that all men of extraordinary strength have sloping shoulders.

A man without sloping shoulders is a man with poor muscular development. Sandow's shoulders slope as much as it is possible for them to do. His neck seems to melt away into his chest. His trapesius muscle, which he fondly loves, is as thick through as the back of a man's hand, as broad, and thicker in some places.

"It will be observed that in some of his pictures Mr. Sandow appears to have a corrugated stomach. This is due to the perfect development of a set of muscles destined to protect the abdomen, but neglected and undeveloped with most of us. On Sandow's stomach these muscles stand out distinctly, each about as big as a man's wrist. He invited his visitors to run their closed knuckles violently up and down this collection of stomach muscles. The effect was that of rubbing the knuckles up and down an old-fashioned washboard.

"From his teeth, with which he can support the weight of a good-sized horse, down to his feet, Sandow is thoroughly developed. Every muscle stands out by itself and appears to be under perfect control.

"His object is to bring out and utilize all the strength that is in him, and his success, which is absolute, makes him beyond question, so far as is known and so far as record goes, the nearest to physical perfection of any living man. It will be a good thing for young men and for boys to study Sandow. It will fill them with ambition to be like him and may add to their wealth, which, in his case, he thoroughly deserves.

"SANDOW'S INCREASING STRENGTH.

"An interesting fact is the constant increase in Sandow's strength. He is very much more powerful now than he was when he went to London a few years ago and easily defeated Samson and his pupil, Cyclops, then reputed to be the two strongest men in the world. That particular contest, which was umpired by the Marquis of Queensberry and Lord De Clifford, attracted one hundred thousand Londoners to the neighbourhood of the Aquarium and packed that institution as it had never been packed before. It is difficult to find in history any man to compare with Sandow, unless one goes back to the far-off days when Samson was edited by Delilah. Thomas Topham, the famous strongman of England, may have been as good a man as Sandow in actual brute strength; but he was very much bigger in build and far less interesting as a demonstration of the possibilities of muscular development. Topham is the man who, according to tradition, pulled successfully against two horses, carried off a sleeping watchman in his sentry-box to leave him in a graveyard, lifted three casks of water at one time weighing eighteen hundred pounds, and lived in terror of a very small wife.

The New York *Herald*, of Sunday, June 18th, also devoted a number of columns to Sandow's advent in the New World, with an interesting, though necessarily brief, account of his career. Having ourselves dealt, in the preceding pages, with the biography, our extract from the *Herald* will be confined within the following brief limits :—

INTERVIEWED BY THE NEW YORK HERALD,

" Perhaps the strongest man," writes the *Herald*, " whom the world has seen since Samson destroyed himself along with three thousand Philistines, is in New York just now. He is not slaying thousands with the jawbone of an ass, carrying off ponderous gates like those of Gaza on his shoulders, nor pulling down stone houses on himself and others, but he is doing feats in lifting dumb-bells, men and horses, that make cold chills chase one another up and down the spine of the beholder.

" Eugene Sandow, this modern marvel of physical power, beside whom the average man is puny, made his American début in the Casino recently before a private gathering of about two hundred persons, many of whom were medical men. It was hard for the spectators, when a calcium light was turned on the figure standing on a pedestal in the back of the darkened stage, to believe that it was indeed flesh and blood that they beheld. Such knots and bunches and layers of muscle they had never before seen other than on the statue of an Achilles, a Discobolus, or the Fighting Gladiator.

"SANDOW'S BOYISH FACE.

"The face was that of little more than a boy—smooth, with rosy cheeks and a little blond moustache. The chin, however, was square and heavy. The neck was massive, and the shoulders seemed a yard apart. The arms looked as though hickory-nuts and walnuts had somehow been forced under the skin, causing it to bulge out in abrupt lumps. Layers of muscle, three inches thick, covered the chest, and on the abdomen was a succession of rolls of muscle that one could tell even from a distance of several yards were hard as iron.

"Sandow's vital organs are undoubtedly as sound as his muscles. The capacity of his lungs is simply wonderful. The popular idea that strong men develop their muscular system at the expense of the vital organs is fallacious. To increase the size of the muscles the circulation must be increased, and this implies, of course, increased work by heart and lungs. The functional capacity of these organs is therefore increased proportionately to the increase of muscle.

"Sandow, in ordinary street dress, gives no indication of the wonderful power he possesses. There are many athletes and oarsmen who look just as strong as he to the casual observer. It is when one touches him or sees him stripped that one gets an idea of his vast strength.

"MUSCLES HARD AS WOOD.

"His muscles, when flexed, are as unyielding to the touch as iron, When Sandow strikes himself on a muscle with his hand, it gives forth a sound like wood. An idea of the size of Sandow's muscles may be gained by the measurement of various parts of his body. The figures, as Sandow gave them to me, are as follows:—

"Neck, $18\frac{1}{2}$ inches; biceps, $19\frac{1}{2}$ inches; forearm, 17 inches; chest, normal, 52 inches; contracted, 46 inches; expanded, 58 inches;

waist, 29 inches; thigh, 26¾ inches; calf, 18 inches; height, 5 feet 8¼ inches; weight, 199 pounds.*

"Sandow does not believe in elaborate training. 'Under my system of getting strong,' said he, 'a man need but follow his ordinary course of life and take reasonably good care of himself. No dietetic regulations are needed. Let him eat and drink whatever suits him. As for sleeping, I don't think it makes any great difference when he sleeps, provided he gets sleep enough. I myself go to bed any time between midnight and three o'clock in the morning. I eat whenever, whatever, and as much as I please. I drink all I can get. Yes; beer, ale, wines, champagne, cognac—everything. But I never drink to excess. I take a very cold water bath every morning and another after my performance at night. Exercise? Yes, a little—what I get in my regular performance.'

"GETTING STRONGER EVERY YEAR.

"Sandow says he is getting stronger every year, and expects to keep increasing in strength for years to come. Before he gives up professional work, he says he will write a book explaining his system. He will also give personal instructions to those who want to become strong.

"The feats which Sandow performs on the stage seem nothing less than marvellous. He handles fifty-six pound dumb-bells as a schoolboy would handle weights of two pounds each. He is not in the least muscle-bound and turns somersaults and handsprings with the ease of a professional acrobat. One of his tricks is to turn a back somersault with his feet tied together, his eyes blindfolded, and a fifty-six pound dumb-bell in each hand.

"In his nightly performance at the Casino, four men carry on the stage an immense dumb-bell, the bar of which is of brass about four feet long, and the bells, which are hollow, three feet in diameter With great effort Sandow raises the bell over his head with one arm, then dropping it suddenly, catches it with both hands and places it

* Mr. Sandow's present measurements are as follows:—Neck, 18 inches; forearm, 16¼ inches; biceps, 19¼ inches; chest relaxed, 40 inches; normal, 47 inches; expanded, 61 inches; waist, 28 inches; thigh, 27 inches; calf, 18 inches; height, 5 feet 8¼ inches; weight, 196 pounds.—EDITOR.

lightly on the floor, whereupon the attendants release a man from each bell. The total weight of the apparatus and men is about three hundred and twenty pounds.

"HOLDING UP THREE HORSES.

"Another feat is that of supporting with his arms and legs the weight of three horses. Sandow rests on his hands and feet with his back towards the floor. A heavy wooden platform is then placed on him, resting on his shoulders, chest, and knees. This platform is constructed to fit about the neck to prevent its slipping or moving in any way. "A long wooden bridge is then placed across the platform, and three trained horses walk upon the bridge. They remain there for about a minute, while every muscle of the giant underneath stands out like whipcord. The weight of the animals and apparatus is said to be 2,600 pounds."

XV.

SANDOW AS A PHYSIOLOGICAL STUDY.

INTEREST in Sandow as a physiological study has always been intense. The finely-formed limbs, the great thews, the Titanic strength, and the splendid heart and lung-power of the famous athlete, have been the admiration of countless medical men and artists in the nude. What has, also, especially struck the medical expert,—and chiefly, the anatomist,—is Sandow's wonderful power of relaxing antagonistic muscles and bringing each into individual play. His facility in this respect is phenomenal, and shows how thorough, and at the same time intelligent, has been his training. A hardly less notable feature in the great athlete is his suppleness of limb and the shapeliness and symmetry of his person. Herein we see the secret of Greek art, as modelled in its famous sculpture, for nature may be trusted to impart physical beauty where the conditions of bodily life and exercise are favourable to the highest forms of human development. Sandow's attraction to those we have referred to, and to all lovers of the strong and the beautiful, may therefore be comprehended, for "creeds decay, scholarship grows musty, and the wisdom of one generation becomes the foolishness of the next; but beauty endures forever."

EXAMINED BY DR. SARGENT, OF HARVARD.

Among those in America who have made Sandow a physiological and anatomical study, is Dr. D. A. Sargent, M. A., the skilled and enthusiastic Director of Gymnastics at Harvard, previously referred to. This eminent authority in athletics, at the request of the New York *World*, made a professional examination of Sandow, and subjected him to a series of elaborate measurements and interesting tests, such as are applied to the Harvard undergraduates entering the Hemenway Gymnasium. Dr. Sargent has courteously permitted the publication in these pages of his report, which is here appended; and we owe our obligations to the *World* for the use we also make of the article which appeared in that journal giving an account of Dr. Sargent's examination.

"The first thing," says Dr. Sargent, "that struck me when I saw Sandow stripped was the extraordinary size of the muscles as compared with that of the bones. His skeleton is not large, as is easily seen in the girth of his wrist and ankles, but the bones are exceedingly fine. The muscles are also of very fine quality. The fibres are unusually small, but they are much more numerous than in the case of the average athlete, a fact which accounts for their great bulk. His muscles in certain regions, notably on the upper arms and back, are developed to an extraordinary degree. The trapezius and extensors and flexors of the legs and thighs are also tremendous. The muscles of the pectoral are not so large relatively as the deltoid, biceps and triceps. This is probably due to the character of the feats he performs every night.

"Another distinguishing characteristic is his voluntary control of his muscles. He can relax and contract them at will, and the fact that he is able to relax antagonizing muscles is a great aid in performing feats of strength. He is able to employ only such muscles as are necessary, and there is thus very little wasted energy.

"He is remarkably well-balanced in temperament. This may be seen in the shape of the head and poise of the features. In this respect he differs from most very strong men. His body is relatively quite long, and his arms and legs relatively shorter. His head comes under what is known as the 80 per cent. class, which shows the possession of the great amount of nervous energy which he throws into his work and enables him to perform his wonderful feats.

"I have found it to be a rule that strong, large men are slow in their movements, and inclined to be dull and stupid. But when you come to put Sandow to the test you find that for a man of his power he is very quick. His time-reaction as shown by the electrical instruments was truly remarkable, and the fact that the speed of his arm in a forward movement was almost equal to that of Mr. Donovan, who is a man of acknowledged agility and with much less muscle than Sandow, is, I think, extraordinary.

"A peculiarity about Sandow in taking a deep breath is that he fills the top of the chest first. You will find it usually the case that a man will naturally begin to fill his lungs at the bottom. But in the machine registering the normal breathing movement it was seen that the abdominal breathing was greater than the thoracic. This is as it should be, though I find it rarely among athletes at Harvard. His breathing is also remarkably synchronous.

"Altogether Sandow is the most wonderful specimen of man I have ever seen. He is strong, active and graceful, combining the characteristics of Apollo, Hercules, and the ideal athlete. There is not the slightest evidence of sham about him. On the contrary, he is just what he pretends to be. His behaviour under the tests was admirable. I might add that he combines with his other qualities those of a perfect gentleman. He has a considerable knowledge of anatomy, and can call the muscles by their proper names. I shall be glad to have him come and lecture before the students of Harvard. It will be a treat for them to see a man of his physical development, and will doubtless act as a stimulus. It is a curious fact that a very strong man always has a host of imitators."

"THE STRONGEST MAN MEASURED.

Herewith is appended the New York *World's* report of the incidents occurring at the examination conducted by Dr. Sargent:—

"By special arrangement with the Sunday *World*, Dr. Sargent, the medical examiner and physical adviser at Harvard University, came to New York last week and made a thorough anatomical test of Sandow, the strongest man in the world. The test was entirely satisfactory. After it was over, Dr. Sargent said that Sandow was everything he said he was, and that he had never before, in all his long experience with Harvard athletes, seen such a wonderfully developed specimen of manhood. The examination was made in a large room in a hotel on Broadway, near Sandow's boarding-house. The room was supplied before Sandow's arrival with a very interesting set of apparatus, designed to test almost every possible exercise of the muscles. There were instruments to blow in, to determine your force of expiration; a machine to find out how many pounds you can lift, another to see how hard you can squeeze, another to register the power of the muscles of the chest, another to measure the exact amount of air you can take into your lungs. There was also an electrical apparatus which was so contrived that it recorded on a cylinder, covered with a thin coating of lampblack, etchings showing how regularly you breathe, and the relative proportion of breathing done by the abdomen to that done by the chest. There were also a set of scales; while Dr. Sargent's secretary, who was present, took down the measurements.

"When Sandow entered the room he had on a suit of steel-gray clothes, with a cut-away coat. Clothing, as a rule, effectually conceals a man's physical development, which is in most cases a fortunate circumstance from an artistic point of view. But it is easy to see that Sandow, even when dressed, possesses marvellous muscular power. His coat bulges out about the chest and back, in curious contrast to the waist, which is as small as a woman's.

"After removing as much clothing as possible, he stood before Dr. Sargent, a fine example of what nature intended man to be. The muscles of his back, arms, legs, and sides stood out in great welts. His finely-moulded head, more like those on ancient statues than you will find in many a day's search, his small waist, and his slender ankles, were in artistic contrast to his wealth of muscle. At this early stage in the proceedings, Dr. Sargent began to be surprised. He was much more surprised later on.

"Sandow was first asked to step on the scales and be weighed. The beam tipped at 180 pounds. This is slightly less than his usual weight, and he attributes the falling off to the recent hot spell. It is interesting to know that this is the exact weight that Dr. Sargent assigns to the typical athlete, a statue of which, constructed on purely scientific measurements, he has sent to the World's Fair. Sandow's height was then found to be 5 feet $8\frac{1}{2}$ inches. The other measurements that coincided with Dr. Sargent's ideal were those of the length of the foot and the girth of the ankles. In all other dimensions, especially those of the muscles on the arms and back, Sandow was considerably larger than the model.

"Among the instruments that Dr. Sargent had provided was an apparatus with two handles fixed to either end of a short steel bar. To this bar was attached a semi-circular plane, with an indicator that moved along a scale, showing the number of kilometers of force exerted when the handles were pressed together. One ambitious person present, after pushing on these handles until he was very red in the face, made the indicator go half-way round. Another gentleman, who is a good deal stronger than one might suppose, made it move around a little further. Sandow then took hold and pressed. The indicator went round until it had passed the last registering mark, and was stopped by a little steel knob. If that hadn't been there the indicator might have described a complete circle. This was one of the features of the examination that especially surprised Dr. Sargent.

"There was another apparatus with an indicator to show how many pounds you can lift. Sandow attacked this until the indicator registered 440 kilos. This is about 1,000 pounds, but Sandow expressed himself as very much disappointed with the result. There was nothing to show for the tremendous amount of muscular power exerted beyond the gradual moving of a little steel arrow along a graduated scale.

"'If you want feats of strength,' he said, 'I will show you something.'

"He then asked for the heaviest man in the room. This proved to be Dr. Sargent himself. He had been weighed earlier in the morning and had tipped the scales at 175 pounds. After expressing his regret that there was no one heavier at hand, Sandow required the doctor to stand with his back towards a table placed in the centre of the room. Sandow knelt down and laid his right hand flat on the floor, with the palm turned up, and asked the doctor to stand on it with one foot. Then, taking a firm hold, he raised the eminent physician rapidly but easily to the top of the table, whence he removed him as gently as a mother would her child. The most remarkable thing about this performance was that the lifting was done with a straight arm. There was not the slightest bending at the elbow. This was another instance at which the doctor was considerably surprised. It was certainly a wonderful feat, and far more impressive as an object lesson than pulling at the machines, though of course that was valuable as a scientific test.

"There was still another machine, which was designed to be placed between the knees and which registered the power of compression of the legs. Sandow was also disappointed with this. He did not take much satisfaction in moving the indicator, no matter how much it registered. So he asked the doctor to sit in a chair opposite him with his knees tight together. Sandow then sat down with his knees pressing against those of the doctor, and told the latter to force his legs apart. Dr. Sargent tugged and strained, but his legs remained locked as in a vise. The situation was reversed, and Sandow pushed the doctor's legs apart as easily as though they had been wisps of hay. As an illustration of his leg power, Sandow said that once an experiment was made in which a horse was hitched to each knee and then started ahead with the purpose of pulling his legs apart. The horses were unable to budge an inch. Sandow then separated his legs and the horses were again started. The knees came together and the horses were pulled back. This would be a difficult story to believe in the case of anybody but Sandow. In the old days, when it was the fashion in England to draw and quarter people for imaginary offences, it is likely that Sandow would have escaped unhurt if he had been subjected to this mode of punishment.

"WONDERFUL ABDOMINAL MUSCLES.

"Sandow afforded another illustration of his wonderful strength, this time selecting the muscles of his abdomen as the means of still further surprising Dr. Sargent. Most persons are not aware that they have muscles on their abdomen, and, in fact, they might as well be without them, for they seldom put them to the use intended by nature, that of protecting the intestines and stomach. On Sandow these muscles are revealed in numerous rolls, which when contracted are very hard, and when you rub your hand up and down them feel like a corrugated iron roof. Dr. Sargent was again called into requisition. Sandow lay down on the floor and asked the doctor to stand on his abdomen. After the doctor had assumed this pedestal, Sandow remained for a moment with the muscles relaxed. Then he suddenly contracted them, and the doctor went shooting up into the air. He said afterwards that that was the first time he had ever jumped from a human spring-board.

"It is usually true of very strong men that they are more or less phlegmatic in their movements. This is accounted for by the fact that one set of muscles often impedes the action of the others. The biceps and triceps, for example, are what are called antagonistic muscles. That is, when one contracts it has to overcome the natural tendency of the other to work in the opposite direction. For this reason big, strong men are often slow in getting about. Sandow, however, is peculiarly constituted. He has the faculty of using only those muscles that are required for a particular motion. When relaxed his arm is as soft as a child's, but when contracted it feels like steel. Dr. Sargent said he had never before seen such remarkable control of the muscles as Sandow has of his. On occasion Sandow can put into prominence any one of the muscles of the body. By a twist of the wrist he can make a muscle appear on the forearm which the ordinary man does not know he possesses. By twisting his head a little, he can make another on the back protrude. He is thoroughly familiar with his own anatomy, and knows all of his parts by their scientific names.

"In still another respect he differs greatly from the average strongman. Dr. Sargent has an apparatus, consisting of a long wooden rule, to which is attached a wire, running parallel with the edge. This wire is divided in the middle, and on either side is a small button, which may be moved along a scale. The object of this device is to see how near you may come to guessing exact distances. On one side of the wire the button is placed half-way between the end and the middle of the rule, and you are asked to arrange the other button a like distance from its end. Sandow did this with wonderful accuracy. In all his attempts he seldom failed to place the button at the right point. This shows that he possesses in a remarkable degree what Dr. Sargent calls the power of perception. In other words, his organism is not merely strong but is fine as well.

"INGENIOUS ELECTRICAL TESTS.

"A series of very interesting tests was made with the electrical machine already mentioned, which registered the quantity and quality of breathing. Two fine needles were made to trace markings on a piece of blackened paper. One of these needles was so arranged that it indicated the breathing done by the abdomen, and the other that done by the chest. The average athlete breathes very little with his abdomen, but the ideal athlete uses it almost altogether.

When the apparatus was attached to Sandow, the needles began a slow up-and-down movement. When he drew in his breath the needles moved up, and when he expelled the air taken into his lungs the needles moved down. Dr. Sargent handed Sandow a paper to read, and asked him to distract his attention as far as possible from his surroundings. Then the spectators gathered about the machine. The upper needle, which accounted for the movements of the chest, rose and fell with a regular movement, making a mark about half an inch long. Meanwhile, the other needle moved as slowly and as regularly, but made marks three times as long. If you observe a dog carefully, you will see that his breathing is apparently done in the abdomen. Sandow breathes very much like a dog, and therefore in the way intended by nature. A woman breathes, ordinarily, chiefly with her chest, owing to the constriction of her clothing. Dr. Sargent says this is injurious, and advises loose waists. The pieces of paper on which Sandow's breathing was registered were afterwards treated with shellac, and will be preserved as an example for students at Harvard.

"SPEED IN DELIVERING A BLOW.

"Among the spectators present, was Mr. Michael Donovan, the instructor of boxing at the New York Athletic Club. Mr. Donovan enjoys the deserved reputation of being one of the most skilful and agile boxers in the country. He can strike a blow with surprising quickness. Therefore, in any test for determining the speed of a forward movement of the arm, he must be a good man who can hold his own with Mr. Donovan. There are very few such. Yet Sandow, with a vastly greater muscular force to overcome, can shoot out his arm almost as rapidly. This fact was determined by means of another electrical apparatus, so arranged that the time taken by the fist in passing through a given distance is accurately measured. It was shown that in sixteen trials the average time occupied by Sandow's fist in passing through a distance of 15 75-100 inches was 11-100 of a second. Donovan's speed in ten trials averaged 8-100 of a second. This is a very small difference. But in a variation of the same test Sandow had the better of Mr. Donovan. A small flag was made to drop by pressing an electric button. A device was arranged to discover the exact interval between the dropping of the flag and the moment when the person undergoing the experiment made up his mind to perform a certain action. The test was precisely the same as in the case of a sprinter, who waits for the falling of a flag or the firing of a pistol to get under way. Out of sixteen trials, it took an average of 22-100 of a second for Sandow to make up his mind. Mr. Donovan's time, under the same circumstances, the average being taken from ten trials, was 23-100 of a second, just 1-100 of a second slower. Sandow's maximum was 26-100, and his minimum 18-100 of a second. Mr. Donovan's maximum was 26-100 and his minimum 15-100. The same experiments were tried with the ringing of an electric bell substituted for the falling of the flag. The results were about the same as in the previous trials.

"When the doctor had finished his tests, Sandow gave a short exhibition for the benefit of the spectators. First, he expelled all the air from his lungs, reducing his chest to its smallest possible girth. Then, after taking a few deep breaths, he filled his lungs to their utmost capacity. The difference in the measurements was fourteen inches. The ordinary big-chested man is proud when he can exhibit an expansion of six inches."

XVI.

SANDOW SPEAKS FOR HIMSELF.

HIS VIEWS ON PHYSICAL TRAINING, DIETING, BATHING, EXERCISING, ETC., ETC.

"SANDOW, as a muscular phenomenon is of comparatively limited interest to the public, save as an exciting, and doubtless engaging, curiosity ; but Sandow, as the culmination of a system which will enable even the weakest to attain a perfect physical development, is an object of stupendous interest to everybody." The above forceful dictum is the shrewd and frankly-phrased judgment of the publishers of this work, expressed in a letter of instructions to the editor on his undertaking his congenial task. The writer takes the liberty to preface this section of the book with the intelligent observation, as it is helpful in indicating the scope and design of what, if we do not fail in our purpose, ought to be the most important and serviceable division of the work. In a matter of such paramount moment, the difficulty is not so much to recognize the importance of the real issue, as to lay the finger precisely upon those forces, physical and temperamental, which, in Mr. Sandow's case, have been at work in the evolution and equipment of the athlete, and have made him the structurally perfect type of man he has become. The inquiry is somewhat simplified by our having to leave out hereditary gifts, of any abnormal kind, among the accounting factors for Mr. Sandow's rare physical attainments and phenomenal strength. A careful inquiry has elicited the fact that Mr. Sandow, as we have elsewhere stated, was no youthful prodigy—physical or mental—and inherited from his parents little beyond a well-made but normal frame, and a healthy but by no means vigorous infantile constitution. What he has become, therefore, is the result of his own earnest, persistent and assiduous training, coupled with a temperamental predisposition to all manner of health-giving exercises, with an æsthetic eye for beauty and grace of physical form. To an innate love of the beautiful and the strong, the influence of education has to be added, in the direction it gave to young Sandow's classical studies, and the ability to appreciate, as was exemplified in his youthful visit to Rome, the manly proportions and rare physical beauty of Old World types of manhood, preserved to us in the painter's canvas or in the chiselled forms of the sculptor's art. The prominence given in his German Fatherland to wrestling and gymnastic sports had also, no doubt, its influence upon the budding athlete, to which, in time, must be added the fostering force and moulding power of habit.

If we seek further for the predisposing causes which led Sandow to attain his high degree of physical perfection, we may find an ingredient, of no mean value, in his great natural capacity for work, especially as a youth, and, in the man, a determination and will-power of undeviating and inflexible purpose. All those things, severally, had their proportionate influence; but nothing told with so much and gratifying effect as ceaseless and hard training,—happily directed on an intelligent physiological basis,—ever stimulated by a lively ambition and an unflagging enthusiasm. Our inquiry, however, will be most satisfactorily met by reference to the renowned athlete himself, aided by such responses as Mr. Sandow has made in interviews with inquiring journalists and reporters in pursuit of their daily or nightly tale of "copy." One of these interviews Mr. Sandow has handed to us, and, in spite of its occasional inconsequential and interrogative form, we take leave to incorporate it in these pages. The interview is as reported for the London edition of the *New York Herald*, for Oct. 5, 1890, from which we copy it.

"A REPORTER'S INTERVIEW WITH SANDOW.

"To see such a man as Sandow is to look on an almost ideal form of muscular development. Statistics of the strength of muscular tissue make it not impossible to believe many extraordinary stories with regard to the feats of strongmen who have lifted 300 lbs. with their teeth and 1,200 lbs. with their hands; but Sandow's one-handed jugglery with dumb-bells weighing over 300 lbs., a 'Roman Windmill' game, in which nearly double that power is exercised, and a proof of his endurance under the fell weight of 2,600 lbs., are performances which knock out all previous records in the same line.

"A natural adaptability for work which will develop the bulk and vigour of the muscles in men who, thanks, mayhap, to hereditary causes, are naturally framed for such exercises, forms but small part of the conditions necessary to success. The important question of training is here of paramount consideration, just as in all other athletic pursuits. The old authority who laid it down that an athlete, to be of any use, should have a comely head, brawny arms and legs, a good wind, and considerable strength, would have more than these requisites in Sandow, who is about middle height—5 ft. 8½ in.—but full-breasted and broad-shouldered beyond all ordinary men, and with thighs and lower limbs of wonderful balance and power. Withal, the young German carries himself gracefully, and might rival in statuesque beauty the Farnese Hercules.

"HOW SANDOW BECAME MUSCULAR.

"It should be of interest to know how such perfect muscular manhood was reached. Had such a man been a very wonderful baby, of great prowess as a boy, or how did it all come about? Has it been due to some super-excellent system of training?

"Sandow, with a smile, remarked to a New York *Herald* representative that he believed as an infant his physique was somewhat above the average, but as this rested on maternal authority only—which is ever the same whatever the baby—it may be taken lightly by the sceptical. In boyish exercises, however, he in time proved himself master of the town. But, granting every natural endowment which might fit mortal for athletic honours, Sandow, now in the flush and prime of manhood, thinks that his present bodily status is due more to training than to natural physical gifts. Not that any amount of muscle culture could possibly bring one person in a thousand to the same pitch of excellence, but that in any particular case the regimen is as necessary as the primal physique on which it is exercised.

"Curiously enough, Sandow is a firm believer in the rational free-and-easy style of living and training which most enlightened modern professors use in preference to the violent methods of older days. Regarding as inimical to health any violent changes in one's habits at any period, Sandow advocates nothing beyond mere temperance in the gratification of every natural desire, the strictest discipline being, in his esteem, not inconsistent with the enjoyment of all the rational pleasures of life. Everywhere the theory of constant light exercise has succeeded the older and heavier methods, and no one more eloquently than our accomplished visitor speaks of the utility of light weights in clubs and dumb-bells, and easy, graceful exercise of all sorts for ordinary practice. All, too, should be done on the ground, as he rigidly insists, and, if possible, under proper supervision of skilled instructors. Sandow himself underwent two years' training at Brussels under a distinguished physician, who had the enthusiasm of an athletic preceptor, tempered by the milder knowledge of the scientific anatomist.

"DEVELOPMENT OF MUSCLES.

" To develop the individual muscles Sandow hit on a system for himself by which every set of fibres in the body receives its due care. For the development of the muscles of the arms, legs, chest and back there are varied exercises, each adopted with a view to getting the maximum of healthy life, and by no forcing means, out of each particular set. This system also worked well in half-a-dozen cases with other men on whom the young athlete subsequently tried it. Innumerable easy and graceful motions, careful avoidance of over-exertion, which interferes seriously with the proper production of energetic growth, and deals grave blows at the health of some of the chief organs of the body; not eating much at a time but regularly, little and often; and a few other simple principles seem the stock-in-trade of the system. But about its practice, no doubt there would be found some more difficulty than in learning to waltz. In repeated interviews with Sandow, by some of the chief lights in the Oxford gymnastic theatres, these theories were accepted as admirable.

"'Did you ever engage in a running or wrestling match?' I asked.

"'Never in any running competition, but I like wrestling better than any other physical pastime. Not a muscle of the body, but it catches hold of and improves: calves, thighs, arms, and back—every little bit of human band and strap—are used. Not only that, but it also does one's wit good. Patience, nerve, endurance, agility, quickness, and coolness are all involved.

"AS GOOD AS THE ANCIENTS.

"'My notion about the ancients—and, remember, their wrestling is just as we have it in all results—is that they were not a bit better men than there are now living, but that occasionally they found a man incomparably better than his fellows. The classical statues are all idealized—the complete dream of the artist who found in individuals some perfect parts, and shaped a form in which no ingenuity could pick a flaw. Of course, a Hercules or Venus may not have been, is not, impossible: in beauty or strength nothing is impossible, but we don't see such men or women everywhere.'

"'You said, Mr. Sandow, that you didn't believe in the rough school of training which fed men on raw meat, etc.?'

"'No; a man should be denied nothing which he desires within certain limits. I never refuse myself anything—I take wine, beer, smoke, and take a turn all round as other men who make the most of life.'

"'Do you know anything about boxing?'

"'Very little; but I practice with friends, though I think professional fighting brutal sport.

"ENAMOURED OF FOOTBALL.

"Of all English games, let me say, I like football best. It is magnificent, not only as a muscular exercise, but it involves at every turn mental strength, coolness, quickness, and judgment. I saw a football match in Lancashire once which beat any other athletic display I ever saw—the men were so bold, swift, skilful, cool.'"

A FURTHER CHAT WITH THE STRONGMAN.

A little more than a year subsequent to this recorded interview, Mr. Sandow was catechised by another reporter, at the athlete's pleasant home in Pimlico, and as we find in the report of it (*vide* "Answers," Dec. 20, 1891), not a little of interest, respecting Mr. Sandow's personal history and mode of life, we make the following further extract:

"The strongman is a young German who speaks English fairly well, good-looking, with light, curly hair, and a fair moustache. He is singularly modest in manner, and our representative could hardly believe that the young Prussian who entered the room (he is not yet twenty-four) was the splendid athlete who defeated Samson at The Aquarium some months ago. In Mr. Sandow's dining-room is a very fine portrait, by Mr. Aubrey Hunt, of the athlete attired as a Roman gladiator, standing in the Colosseum at Rome. The work is an admirable likeness, and shows off the enormous muscles of Mr. Sandow's body to great advantage. £500 has been offered for the painting, but, naturally enough, its owner refuses to part with it.

"'Now, Mr. Sandow,' began the interviewer, 'how is it you have become so strong as you are? Was it by any system of training, or is it a natural gift?'

"'As a child I was not very strong. As a boy at school I became more powerful and muscular than most of my fellow-students. Beyond the ordinary exercise which every German youth goes through, however, there was little training in my case, in the ordinary sense of the term.'

"'What about diet? I have heard that athletes are obliged to obey very severe dietary rules, that they mustn't eat this and they mustn't drink that, until their lives become a positive burden.'

"'Ah! that is not my case,' said Mr. Sandow, laughing; 'I just eat and drink what I want, when I want, and in what quantities I want. Good, wholesome, plain food I find to be best.'

"It was not necessary to ask the young German whether he smoked, for at that moment he was puffing away vigorously at a huge cigar, as if he enjoyed it very much.

"'I usually dine,' he went on, 'about 6:30 P. M., as a rest for thorough digestion is necessary before going through my performance, which, although it occupies only about twenty minutes, is very arduous while it lasts.'

"'What is the greatest weight you raise with one hand?'

"'Over 300 lbs. I lift it from the ground to the head, then to the full length of my arm. It is a much more difficult thing to lift a weight than to support it. Once raised to the head I could sustain almost any burden.'

"Of course you are proficient in most branches of athletics?'

"'Yes, I am a bit of an acrobat in my way. Only as an amateur, however, for I have never appeared professionally in that capacity.'

"'When did you begin to take up your present profession?'

"'Only in the spring of the year after my contest with Samson at The Aquarium. Prior to that I had appeared only as an amateur in Germany, and in a few other countries on the continent.'

"'Now, Mr. Sandow, how tall are you?' our man asked.

"'I am just 5 feet 8¼ inches in height,' was the reply.

"'And how much round the chest?'

"'Forty-eight inches.'

"Forty-eight inches! And the ordinary six-foot guardsman averages only about forty-one inches. This was an astonishment.

"'What does your arm measure round the biceps?'

"'Nineteen inches.'

"The *Answers* man here grasped the athlete's arm. It resembled iron rather than human flesh, and it is just the same all over his body. Nothing but solid adamantine muscle is to be felt, and not one ounce of superfluous flesh is apparent.

"'You don't go in for chain-breaking and wire-rope-snapping feats, do you?'

"'No, I don't care much for them. They are more or less knacks, sometimes mere conjuring feats, indeed, but are, nevertheless, clever.'

"Our representative afterwards had an opportunity of seeing the German athlete go through his performance at the London Pavilion, in company with his pupil, Loris. And, certainly, the feats are wonderful. Nor is there much doubt about their genuineness. The iron weights and dumb-bells are, at the termination of each act, allowed to fall with a very real and solid sound upon the stage, and, moreover, any one among the audience is at perfect liberty to touch and—if he can—lift them. The heaviest weight scales 312 lbs., and this Sandow lifts with apparent ease to his head and holds it there. He can even turn a somersault whilst holding in his hands two 56 lb. weights. He terminates his exhibition by supporting upon his chest, propped by his arms and his legs from below the knees, no less than 2,600 lbs., or over a ton of stone, iron, and human bodies.

"The remark was now made that the athlete would make a splendid wrestler, when the Teuton replied:—

"'In my own country I was a champion, and no one was ever able successfully to contend with me.'

"'What style did you contend in?'

"'The Græco-Roman. That is the only species of wrestling taught in the German Turn Vereins. We know nothing of leg-work, which is the dividing line between the style of the ancients and the Lancashire fashion.'

"'When and how did you first take to gymnastics?'

"'Well, when I was a young man I was a mere stripling, and thought to strengthen my frame by a little light exercise, like the working of a wooden wand or a light iron bar.'

"'Did that do you any good?'

"'Yes; it loosened all my muscles and made them pliant, but no great amount of development came from the exercises. This set me thinking, and I gradually found out what exercises were the best to develop certain kinds of muscles. Using my knowledge with the weights I had at my command, I began to gradually increase my weights and found out that I could easily put up a 100 lb. dumbbell.'

"'How long did it take you to fully develop your strength?'

"'That is a hard question to decide. I do not think that I have fully developed my strength yet; but it took me two years' hard study to find out just where the power came from. Of course, I am finding out new things all the time, and it is quite possible that I may discover some new muscle, which will enable me still further to increase my lifting power.'"

A representative of the same journal (*vide* "Answers," for May 30th, 1893), in a subsequent interview with "the king of strong men," as he has been called, elicited some further information from him respecting training, and specially touching the use of dumb-bells, which we epitomise as follows: Sandow was asked if he approved of the customary drill with the dumb-bells taught in gymnasiums. His answer was an emphatic "No;" "half the motions," he added, "don't affect the muscles a bit, and there are dozens of muscles which are not brought into action at all, and practically lie dormant and untrained. Nor have I much faith in gymnastics as they are usually taught. They don't bring out the muscles one uses in everyday life. Parallel bars and much of the apparatus of training, I have found of little use. My faith is pinned to dumb-bells, and I do all my training with their aid, supplemented by weight-lifting. By the constant use of dumb-bells any man of average strength can bring his muscles to the highest possible development; but he should, of course, know my system, which has been adopted after much careful and scientific study, and has had the approval of the military authorities of Britain, and in the training schools for the army has been put to the most satisfactory tests. If I had a boy," continued Mr. Sandow, "I should start him with ½-lb. dumb-bells when he was two years old, and then gradually increase the weight with his years. My idea is that boys of from ten to twelve should have 3-lb. dumb-bells; from twelve to fifteen, 4-lb.; and from fifteen upwards, I consider 5-lb. dumb-bells quite sufficient for any one. But there is little use, and only a waste of time, in exercising with dumb-bells by fits and starts; they should be used persistently and systematically. It should be compulsory in all schools for boys to have regular training with dumb-bells, and if this were universal there would soon be a most beneficial change in the physique of the rising generation."

The importance to be attached to this evidence of the great athlete, favouring the use of dumb-bells, on his own or any good and intelligent system of exercising with them, can hardly, we venture to think, be open to question. The verdict of a professional, like Mr. Sandow, who has almost solely used them, in attaining the muscular power which enables him to bear the strain of his nightly performances, cannot hastily, at least, be set aside. Nor is there, in his case, mere strength and vigour of muscle, by which he elevates by one hand over his head, a bar-bell, in the bosses of which lurk two men ; supports with ease on his chest a mounted life-guardsman, a grand piano with an orchestra of four men, or the see-saw performances of three good-sized cobs ; there is also that flexibility of frame and suppleness of muscle which enables him, with agility, to turn back-somersaults, with a 56-lb. dumb-bell in each hand, and to carry himself and perform all his movements with litheness and grace.

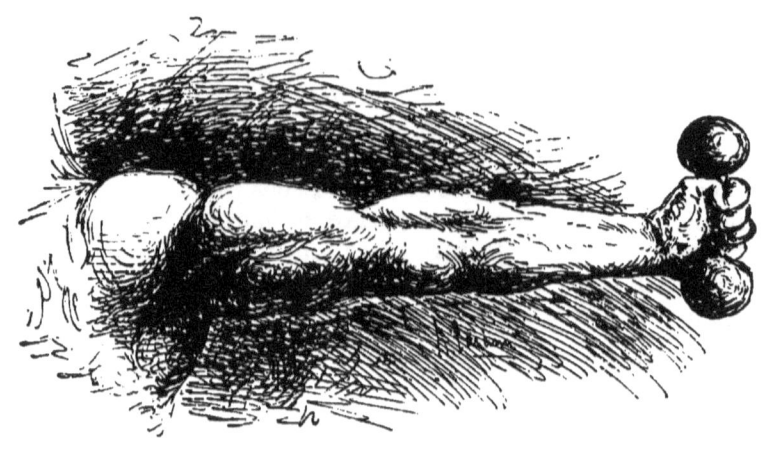

XVII.

THE PHYSIOLOGY OF GYMNASTICS.

MR. SANDOW'S INTRODUCTION TO HIS EXERCISES—HIS VIEWS ON THE THEORETIC AND PRACTICAL BEARING OF PHYSICAL TRAINING.

THE remark has been made by my friend, Colonel Fox, of Aldershot, (see chap. XII.) that I, in my own person, best exemplify the practical results of my system of physical training. Colonel Fox is right, though that gallant officer, at one time, as many others have since, found it hard to believe that my strength has been attained and my muscular system built up by methods of physical training so simple and unambitious as those which I have alone used, and which I commend so earnestly to those who are in search of both health and strength. The disposition is a prevalent one to connect great results with elaborate methods in their achievement, forgetting that Nature does not work so, and unmindful of the fact that the race is not always to the swift nor the battle to the strong. In my own training, I have never made use of elaborate methods, nor, indeed, of any but the simplest; but the latter I have used, as I think, intelligently, and with determination and persistence. It was always an axiom with me to put my mind into my work. Never was there a time when I dawdled over the exercises which experience taught me were helpful in building up my bodily frame and giving me strength and endurance. Nor can I charge myself, at any period of my career, with perfunctorily using the opportunities, whether of time or material, open to me, since I seriously determined to become active and strong. In this, as in everything else, lies much, as we all know, of the secret of success; for without steady application to the work in hand, whether it be to achieve a task or to make oneself strong, the best results are never attainable, if, indeed, we are not likely to make a more or less perilous approach to failure.

INFLUENCE OF BODILY EXERCISE ON THE HUMAN ORGANISM.

Besides concentrating my mind on my work, I have assiduously thought out for myself the best, and, as I have said, the simplest and most effective, modes of training. I have never fancied, nor found need for, the elaborate equipment of the modern gymnasium. Nor have I ever exercised except on the ground, eschewing such appurtenances as the trapezium, the rings, the plank, the ladder, the mast, the vertical pole, and other paraphernalia of gymnastic training. For showy or acrobatic work, these elaborate devices may be useful; but I have not found them helpful as aids to an all-round, vigorous, and healthful bodily development, while practice on them is attended by much and sometimes serious injury and risk.

The dumb-bell and the bar-bell have been my chief means of physical training, aided by a tolerably thorough knowledge of physiology and anatomy, and especially of the ramifications and uses of the muscles. A professional study of the latter, which I was fortunate enough to make at the Medical College at Brussels, has been of very great value to me. It taught me not only the points of localization and functions of the muscles, and their manifold connecting ligatures and tissues, but emphasized, as no desultory or unscientific study could, the physiological effects on the human frame of active and intelligent exercise.

Previous to taking the medical course at Brussels, I had only a layman's shallow acquaintance with the structure and physiology of the animal frame. I had no technical knowledge of organic life, or of the vast field which science, aided by the microscope, has opened up for us in relation to the bone and cell structure, waste and repair, of the human body. I knew little even of the commonest elements of physiology—of the act of breathing and the processes of inspiration and respiration ; of the composition of the blood, its circulation and feeding power ; of digestion and the solvent power of the gastric and pancreatic juices ; of the effect of food, drink, and pure-air on the system. Not only did I learn about these several important matters, which are now, happily, included in an ordinary school education ; but, more valuable to me than all, I acquired a knowledge of the influence of bodily exercise on the human organism—how it affected the blood and its circulation, what its influence was on the organs of movement, on the process of secretion, on respiration, on the nervous system, and especially on the brain, the seat of mental life. These latter studies came to me as a revelation, not only as they furnished me with a guide to health, but as they enabled me to proceed with my muscular training on an intelligent basis, having regard to the just harmony and equilibrium to be preserved in every exercise tending to the building up and strengthening of the human organism.

A SYMMETRICAL AND ALL-ROUND DEVELOPMENT.

Especially helpful were my medical studies in the direction which I sought most for help—to enable me to cultivate a symmetrical and all-round development, rather than a partial and one-sided one. Early in my course, the physiological law, which lies at the bottom of all physical education, was impressed upon me, viz., that the structure and functions of the body increase with use, and that waste comes with disuse and idleness. I then learned to note and appreciate the effect of muscular exercise on the tissue-cells of the body; how bone, muscles, and nerves were affected by muscular action, how it stimulated breathing, and what enriching nutriment it brought to the blood. My next and most important acquisition, was a knowledge of the situation and ramification of the muscles, distinguishing the voluntary and involuntary, as they act dependent upon or independent of the will,—with their uses in the animal economy, in protecting and securing the nutrition of the body, and in carrying on the functions of blood-circulation, respiration, digestion, and secretion. The knowledge I gained of the fibrous net-work of the muscles—the active element in which force resides, and by means of which the bones and joints are brought into play and the movements of the organs are effected,—was, as I have already observed, of incalculable importance to me. It enabled me to locate and bring into due development not only the layers of exterior muscles, the channels of the nerve-force, but the interior muscles also—those which are not seen, yet have active functions to perform, in controlling the movements and aiding the blood-circulation, respiration, etc., within the body. I was not long, of course, in observing the distinctive fact about the muscles—that exercise while it wastes also repairs the body, and, in an especial degree, increases the volume and strength of their own substance; and that muscular action, by accelerating circulation and increasing the absorption of nutritive material, not only assists the regenerative processes of the human organism, but wards off disease and maintains the body in health.

Nor was I slow to discern the advantage to health, physical and mental, in developing, as far as possible, *all* of the muscles; for this is the work which hygienic gymnastics should be made to aim at, if it seeks to secure uniform good health, rather than those purely recreative pastimes, which develop only a special organ or two, to the neglect and disuse of the rest. It is well to impress this, especially on the young reader of these pages or the novice in physical training, for it should not be the mere acquisition of strength, or even skill in the performance of certain feats, that should be aimed at; but that degree of health and vigour of mind, which shall best fit the race for its various vocations, improve its *morale*, and promote its happiness. Especial care, also, should be taken by the young, to see that no exercises are entered upon in excess of the strength of the beginner, otherwise harm, and oftentimes serious harm, will result; nor should they be entered upon in the spirit of reckless and overstrained competition, which not infrequently shortens life or does lasting injury to those engaged in the contest.

EXERCISE SHOULD BE TAKEN WHERE THERE IS FRESH AIR.

Exercise, I would also impress upon the young reader, ought to be taken in a well-ventilated place, not in a contracted bedroom or thronged hall, where the atmosphere is likely either to be close, and therefore poisonous, or contaminated by many breaths, each throwing off at every expiration about twenty cubic inches of impure air, which occasion headaches, laboured breathing, and stagnation of the life-processes. Where exercise is taken in the bedroom, the windows should be open or partially open, particularly if the room be small and the ceiling low. During the exercise, the body should be lightly clad, free from any close-fitting or impeding garment; and, where practicable, a cold plunge bath—even a mere dip in and out—should follow the exercise.

Circumstances will guide those taking daily exercise as to the period of the day in which it is to be indulged. Rest and cessation from work should, for a time, always succeed a meal. The early morning, before breakfast, is best for a little light exercise, or at night before retiring, followed by a bath. In these suggestions, gymnastic exercise for those occupied all day is what I have specially had in view. For walking, running, riding, swimming, rowing, and the active pastimes of the playground, tennis-court and cricket-field, any leisure of the day will, of course, suit, though, in no case, should any considerable exertion immediately follow a meal; and care should be taken that if exercise has been carried so far as to induce perspiration, the body should, if possible, be rubbed dry, and standing in draughts be avoided.

DUMB-BELL AND BAR-BELL EXERCISE RECOMMENDED.

The training I, of course, specially commend is dumb-bell and bar-bell exercise, and, for beginners especially, of very light weights. (For the generality of people, my experience would limit the weight to four or not more than five pounds.) But walking, rowing, skating, swimming, and, where the heart is all right, climbing and running, are very good exercises; football, if not too roughly played, being also excellent. Nothing, in my opinion, however, is better than the use of the dumb-bell, for developing the *whole* system, particularly if it is used intelligently, and with a knowledge of the location and functions of the muscles. With this knowledge, it will surprise most would-be athletes how much can be done with the dumb-bell, and what a range and vast complexity of muscles can by it be brought into play. It has been well said that the muscular system of a man is not made up alone of chest and biceps ; yet to expand the one and enlarge the other is almost all that is thought of by the untrained learner. It is also foolishly supposed that this is the limit of the work to be done by the dumb-bell. Far otherwise is the case, as a subsequent section devoted to exercises will show. The truth is, that there is hardly a muscle that cannot be effectively reached by the system of dumb-bell exercise which I use and have here set forth for the pupil-in-training.

Muscle-culture, of course, should not be taken up spasmodically, or without an object in view, or it will fail of its effect. Nor should the object in view be to develop the muscles merely for adoration or display. Regard ought always to be had to the hygienic benefits to be derived from the exercise. If this be not the purpose of the trainer, the novelty will soon pass and interest will become evanescent. Nor, on the other hand, should gymnastics be pursued violently : prudence should temper ardour and reason restrain recklessness. Perhaps the chief difficulty to be surmounted, especially with beginners who are not young, is to overcome the irksomeness of training and to maintain the interest. Most of us are the creatures of habit, and if physical culture has not been begun early, and been maintained, as it ought to be, through life, new habits, however good in themselves, are difficult to form and pursue with patient assiduity. In this case, the zealous instructor can only fall back on the benefits, mental as well as physical, to be derived from exercise—benefits which are more real than most people are aware of, and are but little understood if muscular exercise is deemed merely a recreation and not a necessity of our being, indispensable to the highest efficiency and health.

INEFFECTIVE AND VICIOUS SYSTEMS OF TRAINING.

In engaging in muscular exercise, or, indeed, in any exercise whatever, much that is beneficial to health is lost for want of an intelligent and well-trained instructor. Even with what is supposed to be a competent instructor, systems of training are frequently adopted that are ineffective, and sometimes vicious. Exercises are not taken up progressively from the simple to the complex. A beginner, at least, should never work in advance of his capacity. Sometimes, too, exercises are indulged in so fatuously as to overstrain the muscles, and, at times, put them to wrong uses. The radical mistake is also made of over-training, and of developing the muscles till they feel like iron, forgetting that flexibility rather than hardness is the symbol and condition of health. Exercise I have, moreover, seen prescribed quite unsuited to the vocation and habits of life of the person counselled to engage in it. Here, as in other things, the old adage is true, that what is one man's meat is another man's poison. The man who taxes his brain all day and leads a sedentary life needs an exercise quite different from that suited to the artisan or mechanic. Both will benefit by a change of occupation, but the brain-worker should have an exercise that animates and exhilarates, and does not fatigue, the mind. For the jaded mind, the best antidotes are sleep and rest.

CORRECT HABITS OF BREATHING.

A further caution to be observed when engaging in muscular exercise, is to acquire correct habits of breathing. In ordinary life few know how to breathe properly, as few know how to sit or stand erect, and maintain, in walking, the proper carriage of body and limbs. When correct attitudes are formed in the bearing of the person, no conscious effort or exertion is needed to maintain them. A careless deportment and slouching poses of the body, so commonly met with, are not only æsthetic defects, but do grave injury to the health, besides retarding, and detracting from, the stature. No better remedy is there for this than the proper training of the muscles, for they are the legitimate props of the frame, and upon them, and not upon the spine and other bone structures of the system, devolve the duty of supporting the body and keeping it erect. If we are to breathe aright, the inflation of the lungs should be from below rather than from the top, that is, that the inspiratory act should fill the lower part of the lungs and diaphragm first, then be inhaled upwards with a lifting and expanding movement of the chest, giving the latter room to distend by throwing back the head and shoulders. Take full, long breaths, and not short, gasping ones, retaining the breath for a time in the lungs and air-passages, so as to distend the ribs and their connecting cartilages, then expel the air slowly and exhaustively, assisted, if need be, by a pressure of the hand on the diaphragm and abdomen. This counsel may appear at first unnecessary, as nothing seems more easy than effortless or natural breathing, and yet few, comparatively, acquire the art of correct, or, what is termed, natural breathing, as singing-masters and voice-cultivators, especially, know to their cost.* But correct habits of breathing

---

* Mr. W. H. Lawton, a well-known tenor of New York, has recently been lecturing on the Art of Breathing, and has very properly laid stress on the use of the diaphragmatic muscle as an aid to good vocalization, in speaking and singing, as well as a means not only of obviating the throat troubles from which many speakers and singers suffer, but also of developing the chest and giving proper poise and perfect symmetry to the body. From a report of Mr. Lawton's lecture this interesting and instructive extract is given:—

'Mr. Lawton claimed that it was not enough simply to direct the student to

are more important in relation to health than as aids merely to the distension and enlargement of the chest. They are of prime value in the duty they have to perform in the mechanism of respiration, by which the blood is purified and enriched. This is the more important for the young athlete to remember, since it is known that all muscular exercise quickens the action of the lungs and the heart, and that by the joint action of these organs there is an augmentation of the life-giving prop-

exhale and inhale forcibly so many times a day. He must be shown how to use the diaphragmatic muscle, he must be told how to expand the ribs, and must learn that the inaction of the abdominal muscles is proof that the lungs are not used properly. When the student is not taught the proper use and control of the muscles necessary in singing, the ribs fail to be raised to the full extent; the chest does not expand sufficiently; less air enters the lungs, consequently less air and less voice are to be expired. Mr. Lawton recommends these exercises in diaphragmatic breathing not only as indispensable to good vocalization, but as health-giving and favourable to a correct and graceful carriage.

"No matter how fine the natural voice may be the singer must learn how to breathe, and thereby how to poise and sustain the voice on the breath. Technically, knowledge of breathing gives free and easy delivery to the production of the tone, enriches the colours, so to speak, of the voice, and perfects the vocal organ in such a manner as to leave no doubt in the minds of critical hearers as to the singer's artistic ability. Singing on the breath is found to be not only the true secret of artistic vocalization, but an important remedial agent in many physical ailments. A society lady in New York attributes her recovery from bronchitis to her lessons in singing, prescribed by a prominent physician.

"The lecturer also referred to the abnormal development of the stomach and abdomen, brought about in great measure by lack of training in breathing. Proper respiration produces erect carriage and this prevents the accumulation of fat and superfluous flesh below the waist. In urging proper respiratory methods, especially for girls and young women, Mr. Lawton points out and emphasises their value as calisthenic and healthful exercises. The lungs are like a sponge. If the walls of the chest prevent the full inflation of the lungs they cannot perform their part in nature's economy. The blood is not properly oxygenated and the vital forces are necessarily weakened. Nature can be aided in this matter, the muscles of the chest strengthened, the chest itself enlarged, and the thorax greatly assisted in its action, and by very simple means. The singer who grows red in the face and the cords of whose neck become painfully tense is evidently little skilled in the art of managing the voice."

erties of the body. The more the breathing is accelerated, the more rapid, moreover, is the throwing off of the waste material in the system and its replacement by new and fresher substance.

With the breathing process carried on properly, with correct habits in the pose and carriage of the body, with plenty of pure air and good wholesome food, much is secured that goes to the founding and maintaining of health. There is but one other chief provision needed for the acquisition and preservation of a healthy body, namely, exercise, and this has been provided in one's own organs of movement. Warmth, it may be said, we have omitted in this enumeration of the body's wants; but warmth, though it is mainly supplied by the food we eat, is largely aided by exercise, for without muscular action not only would heat lack its life-sustaining and energising force, but the nutritive material, which exercise assists to absorb and distribute by means of the circulation of the blood, would be ill-adapted for its great purpose in the animal economy.*

---

* " The employment of the muscles in exercise not only benefits their especial structure, but it acts on the whole system. When the muscles are put in action, the capillary blood-vessels with which they are supplied become more rapidly charged with blood, and active changes take place not only in the muscles, but in all the surrounding tissues. The heart is required to supply more blood, and accordingly beats more rapidly in order to meet the demand. A larger quantity of blood is sent through the lungs, and larger supplies of oxygen are taken in and carried to the various tissues. The oxygen, by combining with the carbon of the blood and the tissues, engenders a larger quantity of heat, which produces an action on the skin, in order that the superfluous warmth may be disposed of. The skin is thus exercised, as it were, and the sudoriparous (perspiratory) and sebaceous (fatty) glands are set at work. The lungs and skin are brought into operation, and the lungs throw off large quantities of carbonic acid, and the skin large quantities of water, containing in solution matters which, if retained, would produce disease in the body. Wherever the blood is sent, changes of a healthful character occur. The brain and the rest of the nervous system are invigorated, the stomach has its powers of digestion improved, and the liver, pancreas, and other organs perform their functions with more vigour. For want of exercise, the constituents of the

The importance of the matters which have been here treated of will perhaps justify a little further dwelling on, before passing to the movements to be hereafter described. What further has to be said will have reference chiefly to the influence of bodily exercise on the frame and the organs of movement ; on the circulation of the blood ; on respiration, secretion, and digestion ; and on the nervous system and the mental life. A later chapter will deal with the muscles, their situation and chief physiological functions.

food which pass into the blood are not oxidized, and products which produce disease are engendered. The introduction of fresh supplies of oxygen induced by exercise oxidizes these products, and renders them harmless. As a rule, those who exercise most in the open air will live the longest.—*Professor Lankester.*

XVIII.

HYGIENIC AND MEDICAL GYMNASTICS.

It needs no emphasising to say here that it is incumbent on every one to conserve, and, as far as one can, increase, to their full development and vigour, his bodily and mental powers. Whatever agents will best promote this, it is admittedly a duty to make use of. One of the chief means for attaining health and strength is, as has been shown, bodily exercise. This, in the main, is within the reach of all; for a trifling outlay can place at one's use, at least, a pair of light dumb-bells, and, in the cause in which we enlist their service, the expenditure of a little time and energy is surely worth the making. Nor is exercise of this kind unsuited to either young or old, for immature limbs can bear, as they will certainly profit by, a modest amount of pleasurable but systematic training; while even old age will feel the invigorating

FIGURE OF ATHLETE.

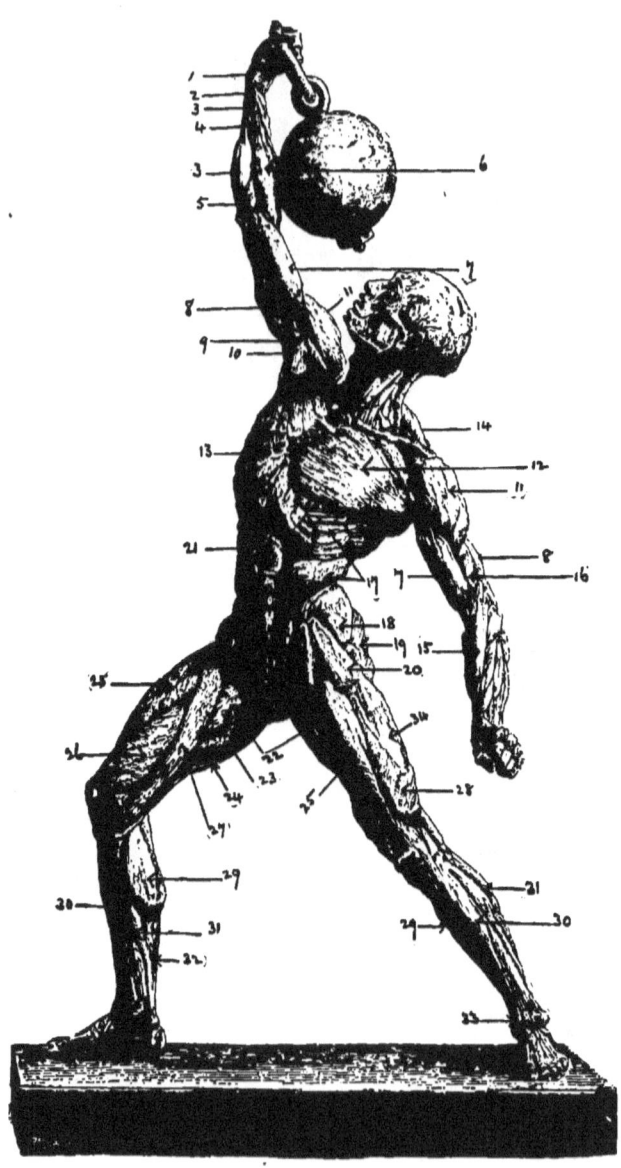

FIGURE OF ATHLETE, SHOWING MUSCLES, ANTERIOR ASPECT.

effects of a little stimulating exercise, which should not, of course, go beyond what is appropriate to declining powers. In the case even of invalids, or of those who suffer from minor and removable ailments, there are many strengthening and curative movements, with or without weights, which would be possible for them to perform, and which would bring relief and perhaps a cure. Of this class, we may mention, among others, those who suffer from chronic headache, rheumatism, indigestion, poorness or imperfect circulation of the blood, nervous troubles, etc., as well as those who are obese or who incline to obesity. For these and such like disorders, a mild course of dumb-bell exercise will be found efficacious, or at any rate salutary; while the exercise can be taken, as if from the home medicine-chest, without resort to the gymnasium or other dispensary.

In truth, the more the rationale of gymnastics is studied, the wider and more beneficial will be the scope of their application as a remedial agent. This is now being admitted by the many who make use of the massage treatment (an agent kindred to gymnastics), and the movement cure. It is also acknowledged by those who pin their faith to dietetics, yet who recognise the difficulty of applying diet-remedies where the condition of the alimentary organs, or any structural weakness of the body, interferes with the absorbing and assimilating of certain foods. Where these defects exist, muscular exercise of a mild character, and when appropriately directed, will be found one of the best means of readjusting the system and furthering the processes of nutrition in the body. Nor is the prescribed remedy inapplicable in the early stages, at least, of consumption and even heart disease, though in these cases, the movements should, of course, be indulged in with moderation. Public speakers and singers will also profit, as we have shown, by exercises which strengthen and give tone to the vocal organs.

In all these several ways can health be promoted, strength acquired, the injurious effect of certain callings in life counteracted, and a very appreciable energising influence exerted upon the mental faculties. To those, especially, whose vocations confine them to sedentary habits and the stooping attitude, and which in too many cases induce distorted frames, round shoulders, or shoulders of unequal height, and a one-sided development of the body and limbs, daily exercise at the dumb-bells will be found fraught with gratifying benefit. But the exercise should be persistent, and, while indulged in, vigorous, for it is unreasonable to expect the frame and its muscle-vesture to recover, by occasional and intermittent corrective exercise, what they are habituated to throughout a long day's occupation in a deforming and unnatural posture. This, it is hardly necessary to say, applies to women as well as to men; for among the other sex are to be met with ill-built and ill-conditioned women, upon whom fashion, unwisely followed, lays its ungracious hand, as seen in the victims of indigestion, constricted breathing, constipation, sallow complexion, the malaise feeling, and feeble health.

EFFECT OF EXERCISE IN BEAUTIFYING WOMEN.

The influence of exercise on the bodily frame of women is, strange to say, still indifferently recognised. The prevalent idea is that muscular exercise of any active kind, for a young girl, coarsens and makes a boy of her. The idea is a delusion; mischievous, indeed, when we realize the value to a growing girl of gymnastics, in their milder form of calisthenics; and its evil results are seen not only in the ailments, among many others, to which we have just referred; but also in the absence of comeliness, grace, and that beauty and shapeliness of physical contour which we associate with a perfectly-formed and finely-conditioned woman. In women, we do not, of course, particularly look for strength, still less for the robust muscle of an Amazon. Nor ought we to look for plumpness only, for a sluggish brain and heavy, inert movements too often mark the merely well-fed but idle woman. It is grace of form and beauty of outline that attract us in the sex, with those genuine accompaniments of efficient physical training—a lustrous eye, a clear skin, a bright intellect, a happy disposition, and a vivacious manner. The antitheses of those charms in a woman—shall I be pardoned for saying it?—are not uncommonly to be met with; but only, it may be said, in one who has neglected the physical need of her nature, and has never known the real joy of living experienced by those who have cultivated the body to a due degree of physical perfection. Only less uncommon are the other physical types of our meagre day—the loutish, half-developed boy, with his lanky limbs and shambling gait, and the gawky girl, with her bony elbows and scraggy neck. Both are culpable human disfigurements whose muscular poverty and general state of ill-condition only parental folly can excuse. Equally lacking in mental and moral wholesomeness must be the boy and girl reared in a reprehensible neglect of physical culture.

The period of youth, I of course remember, is the period of immaturity, and, at an early age, one ought not to look for any abnormal degree of physical development. But I have been speaking of the neglect, not of the paucity, or too-soon-looked-for results, of muscular training. One should be in no hurry to see lads become men and girls become women. Let both be young as long as they possibly can. But youth is the time for laying well the foundation of a sound constitution and the forming of good habits; and the period should not pass, either for girl or for boy, without the salutary stimulus to body and brain of a moderate, regular, and systematic course of physical training. Happily, schools for girls, as well as those for their brothers, are now recognising and supplying this want of adolescent nature, and if the cricket-ground and gymnasium are not yet open to a girl the tennis court is, and she is encouraged to take to rowing, swimming, skating, riding, and mountain-climbing, while wands, rods, and even Indian clubs and dumb-bells, are not the tabooed things they were once to her sex. Were the village green, unluckily, not a thing now of the past, and were corsets and high-heeled shoes not the universal vogue of the time, I should regret the passing away of dancing on the sward. But fashions are mutable, and Newnham and Girton may yet revive the May-pole and its innocent revels, and, at no distant day, it may be, give a degree to a terpsichorean First.

PREJUDICE, INDIFFERENCE, DELUSION.

But I have to do with dumb-bells and, in the main, with exercise for grown or growing men. The complementary exercise to dancing on the sward is wrestling on the green, and if I went into that I should have little space left me for the set purpose of this book. What I want to do here is to endeavour to bring home to every mind the priceless value of exercise on the individual health, and to say a further word or two about the influence which a well-built, healthy body exerts upon the brain. In declaiming on these topics, it is extraordinary to note, in these days of general enlightenment, with what prejudice or indifference the matter is still treated: We educate or cram the mental faculties, often with the veriest lumber in the way of facts, but, partly in the case of one sex, and almost wholly in the case of the other, we leave the bodily powers to take care of themselves. Were the subject of physical training to take its legitimate place among the educating forces of the time, we should startle our school administrators and probably revolutionise society. Both leisure and opportunity would, at least, not be wanting for the proper pursuit of health, of body as well as of mind. Down would go the prejudices, and away would vanish the delusions which now hinder and impede. Indifference also would disappear, and we should no more hear the flimsy pretext of preoccupation or want of time to devote daily to bodily exercise. Of the many delusions which the devotee of physical culture has to meet, there is one I have myself repeatedly had to refute, namely, the assertion that the human body adapts itself as well to a life without, as it does to a life with, exercise. This can be true only in the case of the man who is content to go through life on the lowest plane of vitality. "It is true," as a writer has well observed, "that you may deprive your body of exercise and after a little time you will cease to feel that imperative need of it which a man in perfect health feels if he is by some chance deprived of his accustomed game. But this only means that your body is in a lower condition of vitality. It is perfectly easy to lower the tone of the constitution without being aware of it. The native of a slum in London is certainly less robust than a Yorkshire gamekeeper. But he is not reminded of this fact day by day. He feels the same as usual, and that is all he knows about himself. The questions he ought to ask himself are, what kind of old age is awaiting him? Are his children healthy? If not, is their sickliness to be traced to their father? Lastly, can he do his daily work as efficiently and rapidly as if he were a healthier man?" *

THE BUGBEAR OF TRAINING.

Another, and often a more serious, obstacle one has to contend against, is the want of persistence in exercising, even when you have convinced your friend or pupil of the great benefits to be derived from it. He makes excuse for his de-

* "Health Exhibition Literature," vol. x. p. 128 : London, 1884.

fection on the score of fatigue, and the sense of weariness which the novice in physical training at first feels when he begins to take muscular exercise. This is a trouble which all beginners experience, until the unnatural stiffness or atrophy of the muscles has been overcome and the body yields to the muscular tone which continued exercise in time brings about. A knowledge of the action-processes of the muscles will show what must first take place before the tyro-in-training can find comfort and real pleasure in his work. The muscles, he must understand, work co-ordinately, that is, in a harmonious though antagonistic process, the *flexor*, the bending or doubling-up muscles, situate along the face-front of the trunk and limbs as far down as the knees, pulling in one direction, and the *extensor*, the straightening or opening-out muscles, behind the body and limbs, drawing in another. Hence, until the learner has advanced far enough with his exercises to enable the extensor muscles to respond to his will, and counteract the natural and acquired tendency of the flexors to contract and double-up the limbs, he will feel the effort, and be incommoded by it, to pursue his training with anything like hearty persistence. As time goes on, however, his practising will become easier, and the muscles, at first so much estranged, will act almost automatically and in concert. With increasing exercise will then come not only joy in the work, but such a control of the muscles as will save great expenditure of nervous force, superadded to those gains to health and vital power which the young athlete will find the best rewards of his labour.

HYGIENIC EFFECTS OF EXERCISE.

But it is time, in these talks of the benefits of muscular exercise to health, to turn a little more directly from the popular treatment to the technical; though, in anything I have yet to say, I do not propose, of course, to trench seriously on ground much better covered in the text-books on physiology. I have already, though I fear in a rather desultory way, shown the influence of exercise on the frame and the organs of movement; but something may still helpfully be said of the changes which take place in the muscles as the result of muscular action and its immediate and beneficial effect on the blood. And first it is to be noted that muscular action, aided by the quickened circulation and enrichment of the blood, of which it is a consequence, disintegrates old and constructs new tissue, and by certain chemical changes that take place causes the generation of animal heat. Put in another way, exercise, in conjunction with the quickened action of the heart, accelerates the process of dissimilation, that is, the destruction of the waste material in the body noxious to health, and increases the process of assimilation, or conversion of the food and oxygen into living tissue, the combined process, by the combustion which results, creating heat. This heat raises the temperature of the muscles and enables them to respond more quickly and easily to the behests of the will, acting through the nerve channels, as we see when a person exercising, or otherwise drawing upon his nervous force, warms, as we say, to his work. Secondly, work and heat, thus associated, have a necessary and powerful influence upon the whole life-processes. By quickening the circulation, they hasten the passage of the blood to the heart and from there to the lungs, where, being exposed to the oxygen inhaled it is purified and freshened and set anew on its life-givin mission. A further service which heat renders is to exci the perspiratory glands to do their cleansing and refuse-riding work, aided by the stimulation which is at the same time imparted to their coadjutors—the lungs, the liver, and the kidneys. The more work the skin does in the sweating process, in carrying off the excretions, the less the liver and the kidneys have to do; but health and comfort, in this operation of draining off impurities, demand, as a consequence, proper attention to bathing and frequent ablutions.

MUSCULAR EXERCISE AS AN AID TO DIGESTION.

Thirdly, muscular exercise improves the powers of nutrition and stimulates and strengthens the digestive apparatus. That bodily exercise is a prime factor in promoting digestion and in maintaining the digestive organs in health, is unfortunately not so universally known as it should be. Of the fact, however, there is no manner of doubt. The changes effected before food becomes fit to be taken up in the blood and put in circulation for the sustenance of the body, are, it is true, mainly mechanical and chemical; the former being supplied partly by the teeth and partly by the muscles of the alimentary canal, the latter (the chemical changes) by the saliva, gastric juice, and intestinal secretions. But though this operation is the result of the action of what is termed the "involuntary" muscles,—that is, the muscles which seem to do their work independently of the will,—the "voluntary" muscles, which are specially stimulated by exercise, have an important bearing on the process. Those movements, it may here be said, which specially act upon and strengthen the abdominal muscles are of prime value in their aid to digestion, and should not be neglected by the dyspeptic. By their exercise, not only minor but serious disorders in the digestive organs can be relieved and cured, while a salutary effect can also be exerted on the bowels and intestines, which otherwise not infrequently become torpid. The effect of exercise on the secretions is no less beneficial, for accelerated circulation, it is well known, hastens the gathering-up of the waste matter in the body and its exudation by the great organs of excretion—the skin, the lungs, and the kidneys. Equally vital to the lungs and air-passages is muscular exercise, and the more so if active enough to enlarge the thoracic cavity, or chest, by a full and free play of the breathing power.

In talking of exercise as an aid to digestion, I am constrained to make a slight digression here, that I may speak of the beneficial effects of a plain, wholesome diet in the work of muscle-forming and in giving strength and endurance to the body. On this subject, common as it might be considered among the economies of domestic life, few seem to have any intelligent notion of the nutritive value of foods, or to be able to choose a diet, at once sustaining and palatable, adapted either to one's work or to one's purse. With the lavishness characteristic of the American continent, money is spent like water on the provision for the table, much of which is either unsuited to the system, or injurious to health ; much again is wasted on bad cooking ; while more is thrown away as refuse, instead of being utilised after the manner of the thriftier, yet tasteful and appetising, economies of the French cuisine. It is a pity that there should be such ignorance, wastefulness, and false pride, for all these—if I may be suffered to be so censorious as to say so—are manifested, in too many cases, in the dieting provision and arrangements of American hotels, boarding-houses and households. In these remarks, I have, of course, no desire to air a personal fad, still less to give expression to national prejudice. What has struck me, in the case of American living, is its generousness—a quality which however good in its way, is not always wise in itself, or fairly dealt with by those who are permitted to minister to it. In matters of the table, the popular habit appears to be, to get the best that money can buy, and have lots of it ; forgetting that the dearer meats are often not the most nourishing, and that the plainer foods are the wholesomest, and, where moderately partaken of, are easiest of digestion, as well as the most strengthening. To the sybarite, no doubt, it is pleasantest to consider the palate first, and digestion and nourishment afterwards ; but to the infastidious masses, if their purse does not constrain them, their common sense should, and common sense is not shown in sacrificing nutriment to flavour. Happily, an American expert in physiological matters has recently been taking up this parable and preaching it with intelligent earnestness to, at least, his own people. His arraignment is of the four common mistakes in American households, viz.:—" the use of needlessly expensive kinds of food ; the failure to select the varieties best fitted to our needs for nourishment ; in other words, using relatively too much of certain materials and too little of others ; eating more than is well for health or purse ; and throwing away a great deal of food that ought to be utilised." * What chiefly concerns me in this indictment is the failure to recognise and make use of the food best adapted to the body's wants in the generating of heat and energy. The authority I have quoted is emphatic in affirming that the masses, as a rule, understand little about the nutritive properties of different food materials, as compared with the prices they pay, and with their needs for nourishment. Nor is it a bourgeois taste, but a sound medical judgment, that leads this noted expert to illustrate his argument by declaring that " in buying at ordinary market-rates we get as much material to build up our bodies, repair their waste, and give us strength for work, in 5 cents' worth of flour, or beans, or codfish, as 50c. or $1 will pay for in tenderloin, salmon, or lobsters." He adds that there is as much nutritive value in a pound of wheat flour as in 7 lbs., or $3\frac{1}{2}$ quarts, of oysters, and that, compared with a tenderloin at 50c., a round steak at 15c. a lb. contains as much protein and energy, is just as digestible, and fully as nutritive. To the plutocratic gourmand, who wants to live well, whatever

---

* *Vide* articles in *The Forum* for June, 1892, and Sept., 1893, by Prof. W. O. Atwater, on "Food-wastes in American Households."

the cost, all this may be distasteful and the rankest heresy ; but the wise will probably note the fact, and, it may be, look with more enquiring eye into the ascertained laws of nutrition and the researches of medical men and physiologists interested in health-science and in the laudable economies, as well as the comfort and happiness, of the household. I pass from the subject, content with this simple reference to it, and directing the reader, if he cares to look further into it, to the interesting table in the appendix, on the relative amounts of protein and energy derivable from the different kinds of food. For the privilege of incorporating this table, the publishers are indebted to the courtesy of the Editor of *The Forum*.

From diet to dress the transition is both natural and easy. I have already spoken of the injurious effects on respiration of tight garments, and of the propriety, in taking muscular exercise, of divesting one's self of all restricting and impeding clothing. Much also might be said of the deforming effects, in the case of women and young girls, of tight corsets and small shoes. The necessity for reform in these respects is great, as both are incalculable evils, which may well enlist the ameliorating efforts of those of the sex who earnestly denounce them. With these objectionable things discarded, or structurally modified, so that they will not occasion the ills for which they are now responsible, the health and vigour of women would sensibly improve, the resort to cosmetics would become unnecessary, and the nervous disorders and ailing feeling, which deprive the sex of half the joy of life, would vanish. Then would it be possible for women, whose vital force is now low, to take long walks and indulge in muscular training and derive benefit therefrom ; to be able, with comfort, to trip up and down stairs and feel exhilaration in the exercise ; and to perform all the duties of life with elasticity and freedom. The influence would be no less appreciable in the increased pliancy and grace of the human figure, while, with suitable exercise with light dumb-bells, the contour would improve, and the whole system be toned and invigorated. The effects upon the sterner sex of looser clothing and easy-fitting shoes, with the increased freedom therefrom to take healthful exercise, and with comfort go about their daily avocations, are not the less palpable and real. But the great desideratum is the systematic physical cultivation of the body, and the proper control, as well as exercise, of the muscles. The quickened action of the blood, put in motion by periodic exercise, will do much to dispel the humours which confining pursuits and a sedentary life invite, and give tone and pliancy to the whole organism. By the due awakening of the muscular system, increased flexibility and a higher command of the working machinery of the body will be gained, the joints will be rendered more supple, and either-handedness become as common as it is now rare.

HOW I PASS THE DAY.

Before passing to my concluding topic, the influence of exercise on the nervous system and the mental life, let me say a word about my own diet and training. I am myself no believer in a special diet, still less in a rigid one, as necessary while training. The old nonsense on this subject, about raw eggs and underdone meat, seems to be passing away, and more rational views now prevail. I eat whatever I have a taste for, without stinting myself unduly ; nor do I restrict myself seriously in what I drink. Commonly, I abjure anything intoxicating, confining myself mostly to beer and light wines. Tea and coffee I never suffer myself to touch. All I impose upon my appetites is that they shall be temperately indulged. I endeavour to have my meals at regular hours, and prefer that they shall be simple and easy of digestion. I always take care to chew my food, proper mastication being a *sine qua non* of health. I take plenty of sleep and find this essential to my well-being. As I do not generally get to bed before midnight, or even later, I do not rise until eleven, when I take a cold bath all the year round, preceded by a little light exercise with the dumb-bells. I then have breakfast, and after attending to my correspondence and seeing my friends, I go for a walk or a drive, whatever be the weather. At seven I dine, after which I rest until my evening performance, and close the day with another cold bath and supper. Usually, I dress lightly, though always suitably to the season. My nightly exhibitions, I may add, supply me, together with a good constitutional every day, with all the exercise I need. If I want more, I take it, as I sit reading or smoking, by flicking my muscles.

INFLUENCE OF EXERCISE ON THE MIND.

As an aid to cerebral movements and to the strengthening and clarifying of the mental faculties, no better specific can be suggested than muscular exercise. It is also of great benefit in conserving the nervous force, for the muscular movements have a co-ordinate action on the brain, and seem to stimulate the powers and lessen the fatigues of intellectual effort. Its effect on the nervous system is specially to be noted in the case of those who suffer from fear or timidity, who stammer in their speech, are prone to make grimaces, or lack proper control of the muscles of the face or the body. But its chief importance is the tone it gives to the whole physical system, which enables it to bear the strain incident upon mental concentration, and at the same time to quicken the wit, and render prompt and decisive the judgment. Its moral effect is no less obvious, for it tends to wholesome-mindedness and the tonic bracing of the whole man. For brain-workers, and especially for youth at school and university, where physical education is not assigned its due place in the curriculum, the benefits of copious supplies of good arterial blood is of the deepest importance and should by all means be sought in daily muscular exercise. Without exercise and fresh air, proper oxygenation of the blood cannot take place, and the faculties will lack the invigoration which they ought imperatively to receive. To go on in neglect of this is to stint and impoverish the physical, and to cramp and probably debase the mental, man. The youth at college can have no better zest or stimulant in his studies than an occasional break in their monotony by a little muscular exercise and a restful confab with a roommate or friend. The exercise, at any rate, he ought to have, for without it one can do one's work only under crippling and enfeebling conditions.

Since the foregoing was in type, a thoughtful article on "Child-Study: the Basis of Exact Education," has appeared in *The Forum*, from the pen of Prof. G. Stanley Hall, well-known to readers by his voluminous writings on psychological and educational topics. The article, among other instructive matters bearing on the "natural history of students," emphasises what has been said in the preceding paragraph on the value of muscular exercise as an aid to cerebral movement and to the strengthening and clarifying of the mental faculties. The writer particularly urges, in the case of pupils at school, at a time "when Nature gives man his capital of life-force," increased periods of recreation and improved hygienic conditions under which they shall study, for "work with dulled minds," he forcibly affirms, "breeds all bad mental habits," and, if there is no zestful recreation, no enthusiasm for play as well as for work, passion and self-indulgence will take the place of deep and strong interests in intellectual and moral fields. Muscular education, Dr. Hall insists, ought largely to precede mental training, "especially since thought is coming to be regarded as repressed muscle-action," and since the dry, unrelieved toil and constant tension of school-hours are making great draughts on the nervous system of children, lowering the vital energy, and with it the *morale* and tone of school-life, besides befogging and weakening the brain and inducing all kinds of ailments and disease. The writer admits, of course, that very many children during the school-age would be sickly anyway, and that there are many other causes of sickness besides the school.

"But, on the other hand," he goes on to say, "as shown by many tests, school-house air and bacteria,—even in floor-cracks and in the children's finger-nails—the defective light in some parts of most school-rooms, unphysiological seats, the monotonous strain upon fingers in writing and upon the eye, the necessity of sitting still as the basis of school-work, when activity is the very essence of childhood, the worry of examinations, memory-cram, and bad methods, are, one and all, more or less morbific.
"The modern school is now the most widely extended institution the world has ever seen, and it was never so fast extending as at present. North Africa, New Zealand, Egypt, Finland, and many till lately barbarous lands, under the present colonial policies, have developed elaborate school-systems. The juvenile world now goes to school and has its brain titillated and tattooed, and we have entirely forgotten that men have been not only good citizens but great, who were in idyllic ignorance of even the belauded invention of Cadmus. Now, if this tremendous school-engine, in which everybody believes with a catholic consensus of belief perhaps never before attained, is in the least degree tending to deteriorate mankind physically, it is bad. Knowledge bought at the expense of health, which is wholeness or holiness itself in its higher aspect, is not worth what it costs. Health conditions all the highest joys of life, means full maturity, national prosperity. May we not reverently ask, What shall it profit a child if he gain the whole world of knowledge and lose his health, or what shall he give in exchange for his health?

"That this is coming to be felt is seen in the rapidly growing systems of school-excursions, school-baths, school-gardens, school-lunches, provisions for gymnastics in the various schools, medical inspection, school polyclinics, all of which have lately been repeatedly prescribed and officially normalized. Not all, but many of these are quite new. Here, too, must be placed the interesting tendency to introduce old English sports and even Greek games; the careful psychological study of toys, and the several toy-expositions lately held in Europe; the new hygienic laws concerning school-grounds and buildings, and occasionally books; the rapid growth of vertical script because it requires an erect attitude; new methods of manual and physical training which recognise the difference between the *fundamental*, finer, later and more peripheral *accessory* movements. To select from all these, one, namely medical inspection of schools—this is perhaps nowhere carried farther than in some wards of Paris, where young physicians inspect the eyes, ears, and digestion of each child, and note in a health-book suggestions to both parents and teachers as to diet, regimen, exercise, and studies, besides inspecting the buildings and grounds. The assumption is that all must be judged from the standpoint of health, and that an educational system must make children better and not worse, in health."

CAUTION AGAINST OVER-EXERCISE.

A closing word will not be out of place on the ill-effects of over-straining and unsuitable exercise. The danger in the misuse of athletics is more that against which the young have to guard, for they are apt to misjudge their powers and, in a foolish spirit of rivalry, to over-tax them. This tendency should be frowned upon and discouraged; and to effect this, no number of young men should be permitted to take exercise together, except under watchful and competent supervision. The use, or even the lifting, of heavy weights should also be discouraged, by those at least who have not the adequate strength, or do not know the "knack" in handling them.

Unless well-coached, the young athlete is almost certain to stand badly or poise his body in such a way as to overstrain some muscle upon which the weight should not have fallen, or not have mainly fallen, and, it may be, run the risk of rupture. There is also danger to the heart to be guarded against in indulging in violent exercise. So far as medical testimony goes, however, the instances are not many where injurious results have followed upon even severe physical exertion; and all that need be regarded is to be temperate and sensible. Of course, what may be excessive or unsuitable exercise in one man may not be so in another. Experience and common-sense must here be the judge.

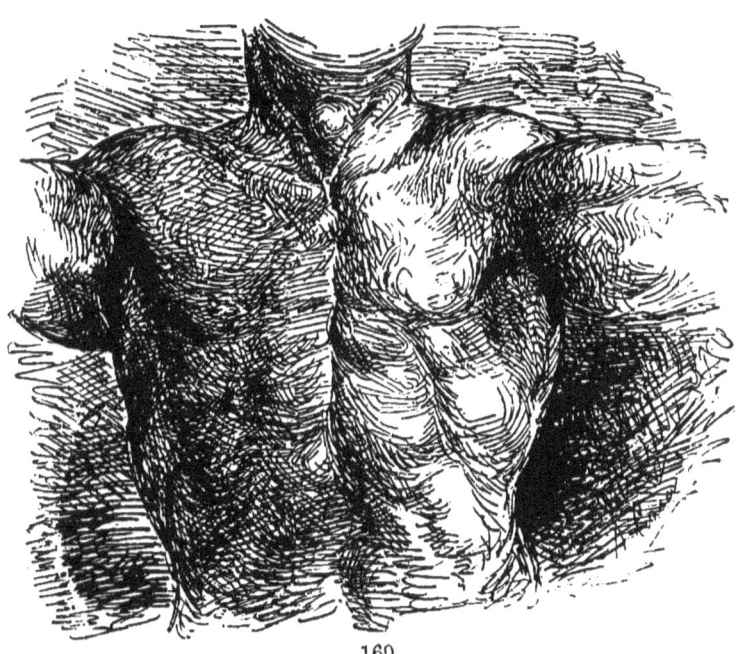

XIX.

EXERCISE AND THE BODILY FUNCTIONS.

MODERN civilisation is seriously discredited by the ignorance usually to be met with in regard to the effects of exercise on the bodily functions. If there were more enlightenment on this subject, it is not too much to say, that the race would live longer and the average health would be higher. Of all topics vital to humanity that of health admittedly is the most important; and yet it is a subject on which few people talk with concern and at the same time with practical intelligence. Converse with any ten men you meet on the subject of physical training as an aid to health, and of the number you will, as a rule, find but one man interested, and, more than likely, he will be a valetudinarian. So long as he is not actually ill, it is extraordinary how content the average man is to go on in almost the lowest plane of vitality, and with

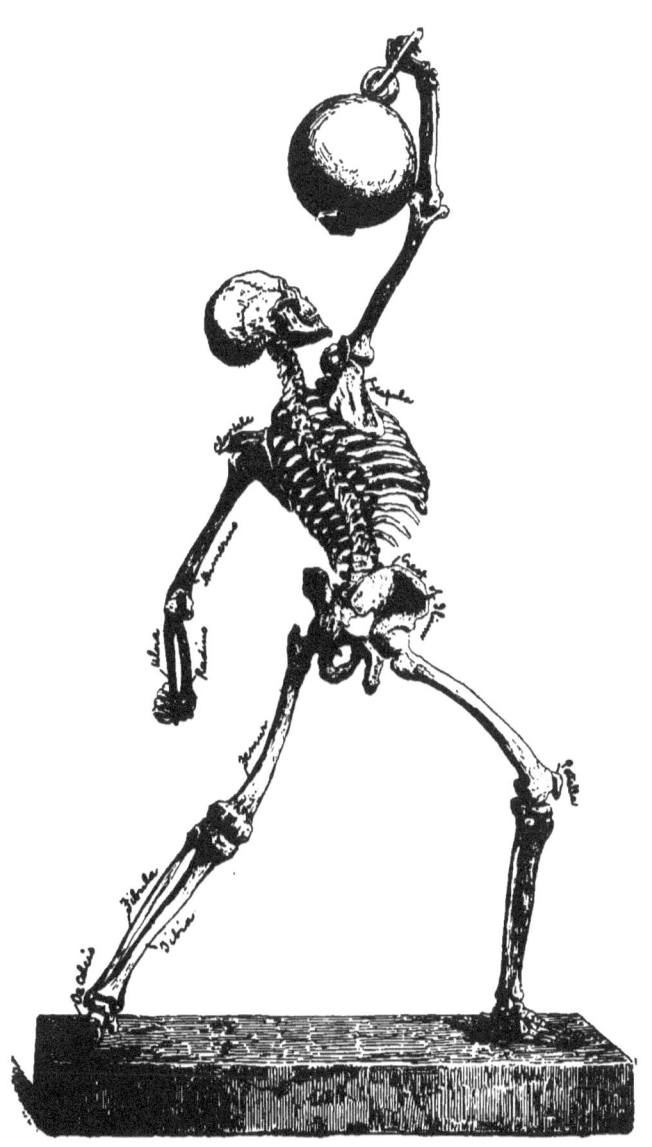

FIGURE OF ATHLETE, SHOWING SKELETON.

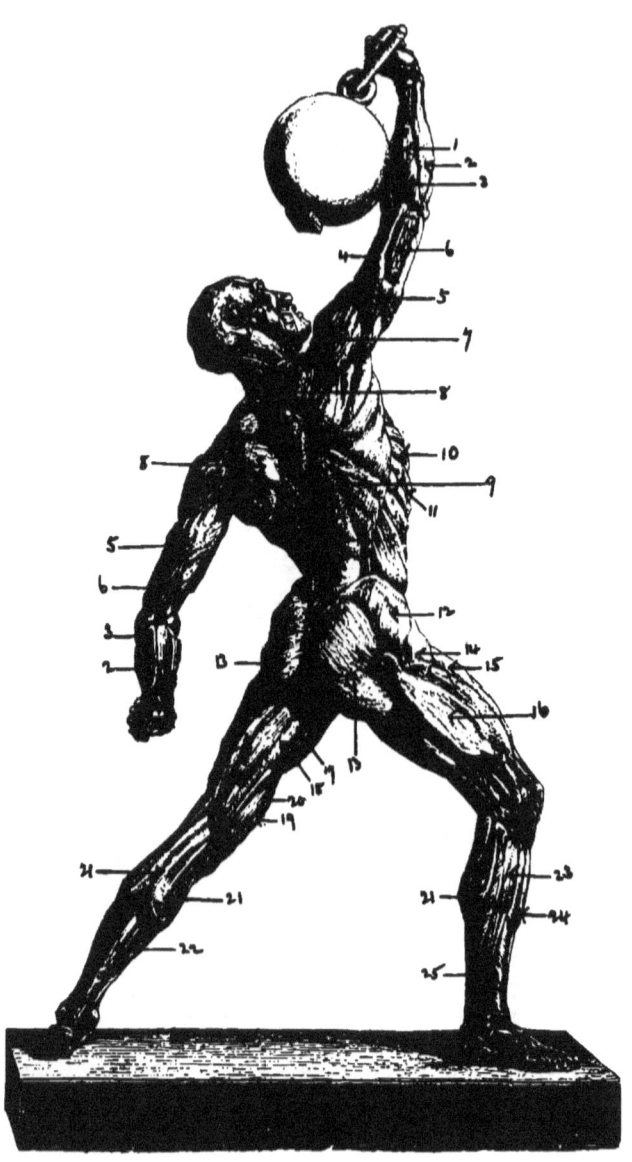

FIGURE OF ATHLETE, SHOWING MUSCLES, POSTERIOR ASPECT.

the minimum both of health and of strength. Nor, when illness finally seizes him, in nine cases out of ten, does he in the least know what to do. In this respect, with all his boasted intelligence, he is usually in a worse plight than his cat or his dog. Without resource in himself, the resort, when he has got tired of ailing, is commonly to the doctor. Then empiricism, more often perhaps than science, has its innings, and, unless he is unusually lucky, he finds that instead of one man not knowing intelligently what to do for him, there are two.

NEGLECT OF EXERCISE AS AN AGENT AND PROMOTER OF HEALTH.

The ignorance we premise on the subject of health and the conditions that best make for it, is, to those who are charged with the public care of it, as startling as it is calamitous. In no province of inquiry is there more pitiful data currently gathered than in that which takes note of the insanitary conditions under which most people live in the neglect of systematic daily exercise as an agent and promoter of health. If we return once more to the subject in this chapter, it is because of its paramount importance, though we do so, we know, at the risk of being charged with advancing only another nostrum for the ailments of the race. Call muscular exercise, however, a nostrum if you will, it can have no kinship with quack remedies in this, that the patient will know what he is taking, and can soon test its efficacy and discern its effects. Given a good system of physical training to work on, and intelligent counsel as a guide, a brief novitiate is all that is needed to produce results that will astonish as well as gratify the most sceptical.

THE AMBITION COMMENDABLE TO BE HEALTHY AND STRONG.

A course of training entered upon with the design of promoting health, and, so far as one can, of perfecting the human organism, is surely worthy of more than a lukewarm interest. The ambition to be healthy is no less commendable than is the ambition to be successful, skilful, or strong. Health is axiomatically affirmed to be the first of requisites, yet how much, and again how little, do we severally mean by the saying? The life-preserving instinct still survives in the race, but in myriads of instances, from our ignorance or contumacy, how little is that life worth living. Our whole manner of life is now a constant disregard of healthy instincts and a crass setting of nature at defiance. "We have perfected every mechanical invention," observes a thoughtful writer, "while we have suffered our bodies—the most perfect machine of all—to atrophy or rust." For specimens of exultant health, we perforce have nowadays to go to the savage. Our cities, with their artificial life and acres of contracted fusty flats and miasmatic tenement houses, do not produce them; nor hardly do we dare to look for them even in the country, where the feverish excitements and degenerating conventions of the town have now penetrated. So far from seeking in these quarters for sound bodies and robust health, we have come to look for, if not wide-spread disease, the conditions that but too surely make for it. On every side is seen a criminal disregard of the physiological laws of health, and, as a consequence, all kinds of physical and mental disorder, with an unarrested and well-nigh unregretted decadence in the higher functions of the human body. Instead of aiming to live, as we might, a joyous healthy life, unchequered by the penalties we must pay for our physiological sins, we have come to regard our everyday and all but universal ailments as the normal condition of mundane existence.

"The farther you have strayed from Nature," observes an able medical writer, "the longer it will take you to retrace your steps." The remark reads like a satire on the dismal efforts of our moral regenerators to *improve* upon Nature, to counteract the vicious tendencies of modern life, and to do everything but stay at its source the progress of physical degeneracy. The same authority we have cited adds, with a touch of pardonable cynicism, that "we have countless benevolent institutions for the prevention of outright death, but not one benevolent enough to make life worth living." Could we have in every town free gymnasia as we have in many free public baths, the reproach, in large measure, might be removed. But we have, as a nation, grown reckless of the public health, as we have grown callous in respect to its claims. This the mortality tables of our large cities, with their appalling record of the march to the grave of half-spent lives, seem distressingly to prove. Our supineness, we suppose, will go on until either some commanding voice arises to prick effectually the stifled conscience of a heedless, though humane, people, or a time of national peril will again come, when the physical vigour of the nation's muscle-defence will be tried in the balance and found wanting.

THE INTER-RELATION OF BODY AND BRAIN.

Did we give heed to the subject, there would be no doubt of the supreme value of daily muscular exercise to the mental and bodily system. We bracket the two, for physiological science has put beyond question the inter-relation of body and brain, and the great activity of function which results from the expenditure of muscular energy. It is one of the interesting points in the study of this subject, to note the physiological effects of exercise on the human organism. Alike in the brain that thinks and in the muscle that acts, results are immediately visible which are as striking as they are incontestable. Negatively, this is shown in the identical disturbances that take place in the system after excessive physical exercise, and after exhausting mental toil. Physicians tell us, among other things in common, that there is the same turbidity of the urine, due to the imperfect burning of the nitrogenous waste substances, which have otherwise to be eliminated from the system. The points of positive similarity are no less remarkable where the results are stimulating and beneficial. There is the same increase in the blood-supply after exercise, and a greater production of heat, both being essential to strenuous bodily and intellectual effort. The fortifying and invigorating influence of active blood-circulation every one has experienced for himself : when the muscles are heated, the functional activity increases, and the body is then most capable of energetic action. Under active exercise, to use an ordinary figure, we increase the fire-draught, and with increased fire-draught there is more rapid combustion and therefore more heat. This functional stimulation of the body necessarily calls for greater supplies of oxygen, and this again produces enrichment of the blood, and has an energising effect upon the nutritive processes. The heart, moreover, undergoes change of size and structure, frees itself from impeding fats, and becomes more fitted for its arduous work. With exercise which increases the contractile power of the muscles, the muscles themselves become more elastic, less susceptible to injury and fatigue, and more firm and enduring. Thus, like the workman who has command of his resources, and can improve the tools of his craft, he who, by exercise, keeps his body in good form is best able to use his organs of work and movement to the fullest advantage and likely to maintain himself in the highest degree of health. Practice will, at the same time, teach him the best methods of utilising his forces —how to economise his breath, conserve his strength, and call to his aid the muscles most fitted for his daily tasks.

MR. SANDOW REMARKABLE AS A HUMAN MOTOR.

Habituation to exercise not only renders hard work easier to perform, but it economises the effort necessary to accomplish it. Mr. Sandow is himself a striking example in this respect; you never see him either breathless or excited, and, even under severe strain, his heart-beat is very uniform, and seldom does he perspire. Beyond any one we have ever seen, he has the most perfect command of his powers. Not only are his muscles marvellously strong, but, to a phenomenal degree, he has acquired the knack of intelligently using them. In the novice, the action of certain muscles is paralysed, as it has been phrased, by the awkward intervention of their antagonists. It is not so, we need hardly say, with the renowned athlete. Every muscle, in his case, is so perfectly trained, as well as under such immediate control, that it does its own assigned duty; while co-ordination of movements is with him rigidly yet unconsciously practised. The effect of this is to distribute the burden of the heavy weights he supports evenly over the muscles, so that no one of them is put to an undue strain. As a human motor, there is not only wonderful strength in Mr. Sandow's muscles, but remarkable facility and ease in their working, amounting almost to automatism; while there is little or no drain upon the nervous system.

THE SECRET OF HEAVY-WEIGHT LIFTING.

In the case of notable athletes, the chief secret of being able to bear great burdens is this, that they know how to distribute the strain of the heavy weights they lift over the whole organism, calling into aid not only the muscles of the arm, but those of the trunk and legs, as well as utilising the main framework of the body, the vertebral column, pelvis, and bones of the lower limbs. They have also learnt the art of so poising the frame that any heavy weight held aloft by the arm shall be parallel to the general direction of the vertebral column, resting upon the nicely-balanced lower limbs and the firmly-planted feet. The co-operation of the bones and muscles of the whole body becomes with practice so easy, that the movements they engage in are accomplished almost automatically, and without taking possession of the brain, or, as we have said, consciously drawing upon the nervous force. That this can be done at all, is one of the curious facts in mental science, for the spinal cord, which is primarily a conductor of movements initiated by the brain, seems to have a memory, and, after a certain habituation to the work to be performed, is able to repeat the movements without much, if any, intervention of the will. Fatigue thus becomes a muscular, rather than a nervous strain, a matter of prime importance to the athlete.

THE PROBLEM OF OBESITY SOLVED.

The absence of fat in the human machine is another of the advantages to the athlete, as it prevents clogging of the muscles and the breathlessness which a fat man suffers from by the formation in the system of carbonic acid, caused by the rapid combustion of the fatty tissues under active muscular exercise. For the reduction of fat, as well as for producing that perfect equilibrium of the functions most favourable to health, there is no better specific than systematic physical training. With persistence in training, and especially in performing the exercise, No. 15, prescribed in the practical section of this volume, the problem of obesity can be solved, aided by the usual precautions as to diet. The exercise to which we have referred is important in this, that it effectively attacks the constitutional as well as the reserve fat tissue, in the region which has an awkward tendency to conserve it, and if constantly practised will reduce its extent, if it does not cause it wholly to disappear. The elimination of the fat will get rid also of breathlessness and the excessive aqueous secretions induced by active exercise. The excretory organs, moreover, will have less to do, and, with advantage to the health, be more free to assist the process of dissimilation.

We need hardly do more than mention the necessity of the daily cold bath after exercise and plenty of fresh air while exercising. We shall, later on, have more to say of these essentials; meanwhile, their importance should not be overlooked by the athlete-in-training. The skin has functions to perform, excretory as well as respiratory; and it is of vital consequence that it should be enabled to do its dual work under the most favouring circumstances. Not less essential to the bodily health and vigour, is the need of copious supplies of pure vivifying air, if the blood is to be well-oxygenated and vital activity promoted through respiratory energy. Good nourishing food and abundant sleep, with, if practicable, occasional intervals of repose during the working hours, should not be neglected, while the physical as well as the mental man will be the gainer by maintaining, as far as possible, a tranquil and unharassed mind. Attention to these, among other points elsewhere dwelt upon, will be of importance, especially to the youth who seeks in systematic muscular exercise to improve his bodily functions and maintain himself in robust physical health and active mental vigour.

XX.

THE CHIEF MUSCLES, WHERE THEY ARE SITUATED, AND WHAT THEY DO.

It will be convenient if we devote a page or two to a brief description and naming of the chief muscles actively concerned in the movements of the body, or of parts of it, so that we may know them when they are designated in the exercises, and apprehend their functions. In the human body, the muscles are of two kinds: (1) those that belong to the animal life, named the *voluntary* muscles, as they act in response to the will; and (2) those that are concerned with the organic life, named the *involuntary* muscles, which, as a rule, are not controlled by the will. The former furnish the machinery of locomotion and work, by the use of which we perform all the acts of life, as in walking, running, lifting, carrying, breathing, speaking, singing, etc. The latter subserve the purposes of organic life, and have important functions as aids to nutrition, digestion, circulation, etc. The two kinds are otherwise distinguished as striped and unstriped muscles, each varying somewhat in its structure, the striped being, as a rule, fibrous and striated, the unstriped smooth and cellular. Both are endowed with the property of contractility, the voluntary muscles contracting more rapidly than the involuntary, and markedly so as the result of active bodily exercise. The voluntary muscles, as they lie chiefly on the surface, form, with the skin, the protective sheathing of the body, and are the means by which the bones are fastened together and made to hinge on their joints. Their form is generally either flat and ribbon-shaped, or bunched up in the middle in short layers, with tapering ends attached by sinews or tendons, at the one end to a fixed bone, designated the muscle-origin, at the other to a movable bone, or integument, designated the muscle-insertion. The voluntary muscles have this peculiarity incident to their contracting power, viz., that they are, for the most part, so arranged as to antagonise or oppose each other, one set pulling in one direction, the other set pulling in another. This contrariety of action we see in operation when we open and close the hand or bend and straighten out the arm. The muscles of the head, the shoulder, the back, and the lower limbs, act in the same way, there being for every motion in one direction a counter-motion in another. In the case of the involuntary muscles, it is this contracting power that operates on the blood-vessels and the intestines, by forcing on, in the former, the circulation of the blood, and the passage, in the latter, of nutritive or excretory matter through or out of the system. The two sets of antagonistic muscles are named, as we have hitherto indicated, the *flexors*, or pulling and drawing-up muscles, and the *extensors*, or relaxing and opening-out ones. The operation of the flexors is seen in the arm when it is flexed, that is, bent or pulled up. It is also seen in the palm of the hand when the fingers are closed, and in the lower limbs, when they are drawn up at the doubling of the knee-joint. The action of the extensors is seen in the complementary motion, which is the reverse of all these, as in the operation of the triceps, at the back of the upper arm, which extends or straightens the forearm out when it has been flexed or doubled up. Another instance may be given of the counteraction of opposing muscles, namely, in the case of the deltoid muscle, the thick, fan-shaped, fibrous layer which envelops the shoulder, and whose function it is to raise the arm—set against the pectoral muscle (the Pectoralis Major) which covers the upper and forepart of the chest, acting in concert with the lumbar and dorsal muscles (the Latissimus dorsi and the Teres Major), whose combined function it is to draw the arm down, assisted, of course, by its own weight.

Generally speaking, the voluntary muscles only are acted upon by the will, communicated through the motor nerves. Exercise, we have already seen, stimulates the action of the muscles, for muscular contraction produces animal heat, and a heated muscle, we know, acts more quickly and powerfully than one at a normal temperature. Heat, moreover, quickens the action of the heart, and this again has its effect on the blood, and, through the blood, a prime factor is set in motion in repairing the waste material and renewing the life-giving properties of the body. The more actively we call the muscles into play, the more beneficial will the results be on the strength and health of the body. But the muscles are not only the vehicles through which the will acts, in inciting to labour and movement; they play an important part in the functions of breathing, speaking, seeing, and hearing. They also perform an involuntary service in giving expression to the feelings and emotions, through the muscles of the face, including those of the eyes and the mouth. In this latter respect, healthful exercise becomes, one might almost say, a moral duty ; for it not only lightens up the face and gives mobility to the muscles of expression, but has a bracing effect on the mind and an enlivening influence on the spirits. Nor should we forget that the muscles form more than one-half of the bulk of the body, and if we neglect their due development and withhold the invigorating influence which exercise exerts upon them, we commit a crime the gravity of which few adequately appreciate.

The muscles, of which there are at least five hundred in number, are named from their uses, shape, situation, and direction ; sometimes also from their points of attachment, as well as from the number of their divisions ; such as the Biceps and the Triceps, the two-headed flexor, and the three-headed extensor, muscles of the arm. We shall confine our notice only to the chief voluntary muscles, to which the subsequent course of exercises will in part refer, and for the development and strengthening of which the movements are especially designed to aid. In the following enumeration and memorandum of the functions of the muscles treated of—namely, those of the upper and lower limbs, the thoracic cavity, and the trunk—we owe our indebtedness chiefly to the great text-books on anatomy, of Quain and Gray. Our descriptions will be materially helped by the anatomical plates reproduced in these pages, after drawings by Prof. Roth.

Let us first deal with the muscles of the upper and fore part of the chest, the shoulder, the arm and forearm ; being those that come into play as the chief organs of motion, and, with the intercostals, the great muscle of the lateral thoracic region—the Serratus Magnus—and the pectoral muscles of the anterior thoracic region, assist in the process of breathing, among their other important functions. The first place has to be given to the DELTOID muscle, which covers the shoulder, whose function it is to raise the arm directly from the side, so as to bring it at right angles with the body. Its fore-fibres, aided by the

Pectoralis Major, the broad triangular muscle situate at the upper and forepart of the chest, draw the arm forwards and upwards, while its rear fibres, assisted by the Teres Major and Latissimus dorsi—the muscles that extend from under the shoulder-blade over the lumbar and lower half of the dorsal regions—draw it backwards and downwards, or enable it to rotate when extended. The deltoid acts as a cap and protector to the deep structures of the shoulder-joint, its muscular fibres being coarse, and so disposed in layers as to reinforce one another and increase their functional power. For its power, it depends mainly upon the shoulder-blade, steadied by the serratus magnus, the head of the triceps, and the middle fibres of the trapezius muscles, to be hereafter described. When the deltoid has raised the arm to the horizontal position, its further elevation is effected by the serratus magnus and trapezius. The rounded prominence of the shoulder is due in part to the thick coating of the deltoid, but mainly to the form of the upper extremity of the arm-bone, which can be felt moving under the muscle as the arm is rotated.

The PECTORALIS MAJOR, which adjoins the Deltoid muscle and has its attachment below it, extends from the region of the collar-bone in front of the armpit over the anterior portion of the ribs. We have already pointed out one of its functions, in conjunction with the Deltoid muscle; another enables it to draw the arm forward and rotate it inwards upon the chest. It also performs another service, in concert with the Pectoralis Minor and the Subclavius muscles, which lie beneath the Pectoralis Major, in drawing the ribs upwards and expanding the chest, when the arms and shoulders are fixed; and is thus an important agent in forced inspiration. Another important muscle, allied with the Pectoral and Subclavius in the latter work, of elevating the ribs and dilating the chest, is the SERRATUS MAGNUS. This muscle, which forms the inner wall of the armpit, wraps the eight upper ribs on the sides of the chest, its deep surface resting upon them and the intercostal muscles. It assists the Trapezius muscle, which covers the upper and back part of the neck and shoulders, in supporting weights on the shoulder. It also lends its aid as a muscle of forced inspiration. The serratus magnus muscle, by withdrawing the base and lower angle of the shoulder-blade from the spinal column, enables the arm, when raised from the shoulder, to be still further outstretched, as in lunging with the dumb-bells. It also comes powerfully into action in all movements of pushing.

The TRAPESIUS is the flat, double muscle, triangular in form, which covers the upper and back part of the neck and shoulders, extending from the posterior part of the head, and from the spinal column in the neck and back to the back part of the collar-bone and shoulder-blade. It is divided in two by the upper part of the spine. The function of each, separately, is to draw back the shoulder-blade, and by rotating it, to raise the shoulder, and also to draw the head and spine to the side. Jointly, the two trapezii have power to pull the head back and to draw the shoulder-blades towards the spine. The LATISSIMUS DORSI is the broad, flat muscle which covers the lumbar and lower dorsal regions, beneath and below the trapezius. It extends obliquely upwards on both sides of the spine from the dorsal vertebræ, where it has its origin, to its insertion in the inner and front portion of the humerus, or arm-bone, near the upper end, and not far from the shoulder-joint. Its action, in concert with the pectoral muscles, is to draw the arms inwards and backwards, as in the act of swimming; or, if the arms are extended or elevated, the latissimus dorsi has the power, jointly with the pectoral and abdominal muscles, to draw the body forwards, as in the act of walking on crutches, or upwards, as in climbing.

We now come to the muscles of the **arm and forearm,**

which have so much to do in the active and useful work of life. For this work, the upper-arm is especially strong, being furnished with long, stout muscles, bunched in the middle by deep layers of connective tissue, and firmly fastened in their end-settings so as to withstand the strain they have to bear. On the front face of the arm is the two-headed muscle, the BICEPS, which is so prominent as to be seen and felt, from its origin, at the head of the arm-bone, underneath the triangular point of the deltoid muscle which envelops the shoulder, to its flattened tendon insertion near to and below the bend of the elbow. This long spindle-shaped muscle, which occupies the whole of the front surface of the upper arm, is the great flexor of the arm ; it also acts as a supinator, and serves to render tense the fascia, or small membranes, of the forearm, by means of the broad enswathing band given off from its tendon. When the forearm is fixed, the BICEPS and BRACHIALIS ANTICUS flex the arm upon the forearm, as seen in the effort of climbing. On the hinder part of the arm, extending along the entire length of the arm-bone, is the TRICEPS muscle, the sole extensor, or straightening-out, muscle of the forearm. In this capacity, its action is that of a force applied to a lever of the first order. The triceps is the direct antagoniser of the biceps and brachialis anticus, for when the latter muscle bends the arm upwards at the elbow, the former draws it into a right line with the arm again. The BRACHIALIS ANTICUS is the muscle which lies immediately behind and projects on each side of the biceps. It covers and forms an important defence to the elbow-joint and the lower half of the front arm-bone, and is, as we have pointed out, a flexor of the elbow. The CORACO-BRACHIALIS, the remaining muscle of the front upper-arm, is the small slender one arising in common with the short head of the biceps, from a process of the shoulder-blade extending down to the middle of the inner side of the humerus, or arm-bone. Its action is to

MUSCLES OF THE FLEXED ARM.
ANTERIOR, POSTERIOR AND LATERAL ASPECTS.

PLATE V.

MUSCLES OF THE TRUNK, SHOULDER, EXTENDED ARMS AND FLEXED

draw the arm forwards and inwards upon the side of the chest.

The muscles of the forearm, though more numerous, need not especially detain us. They are divided into two groups, the front or inner, and the rear or outer groups, each again being divided into the surface muscles, and those of the deep-lying layers. As will be seen from the drawings that illustrate the region, most of the muscles have their source in the upper arm and enfold and protect the two bones, of which the forearm is composed—the ulna, or elbow-bone, the larger of the two, and the radius, or outer bone, lying parallel with its fellow and reaching from the elbow to the wrist. The front or inner group consists of the flexors and pronators, that is, those that bend or turn the forearm, wrist, and hand ; the rear or outer group comprises the extensors and supinators, the direct antagonisers of the front group and that pull in the opposite direction. Most of the muscles of the forearm not only have, as we have said, their source in the arm proper, but are considerably strengthened by the tendinous fibres derived from the three great muscles of that upper limb. Of the flexor and pronator group, there are five superficial and three interior or deep-lying muscles. Technically, their actions are described by Gray as follows :—" Those acting on the forearm are the pronator radii teres and pronator quadratus, which rotate the radius bone upon the ulna, rendering the hand prone (that is, turning the palm downward) ; when pronation has been fully effected the pronator radii teres assists the other muscles in flexing the forearm. The flexors of the wrist are the flexor carpi ulnaris and radialis, and the flexors of the phalanges (the line of the small bones of the fingers) are the flexors sublimis and profundus digitorum ; the former flexing the second phalanges, and the latter the last. The flexor longus pollicis flexes the last phalanx of the thumb. The three latter muscles, after flexing the phalanges, by continuing their action, act upon the wrist, assisting the ordinary flexors of this joint ; and all those which are attached to the humerus (arm-bone) assist in flexing the forearm upon the arm."

The muscles that form the outer and rear group of the forearm, are three in the first division (the Radial region), and four surface and five deep-lying muscles in the second division (the Posterior brachial region). These, the antagonisers, as we have said, of the flexors, comprise all the extensor and supinator muscles, that is, those that straighten and turn upwards the forearm, wrist, and hand. One of the latter, the Anconeus, situate behind and below the elbow-joint, assists the triceps, of which it is a continuation, in extending the forearm. Others act in extending the wrist, etc. ; while still others do duty in turning the arm, wrist and hand upwards. For an enumeration of the intersecting muscles of the wrist and hand, we must refer the reader to the anatomical manuals. A glance, however, may be permitted us at the chief muscles of the abdomen, and those of the lower limbs and the deeper layer of the muscles of the back. And here it may be proper to remark, though the fact is generally applicable, that the force exerted by any muscle during its contraction is in proportion to the number of muscular elements or fibres composing the muscle. This statement is made on the warrant of Quain, the famous anatomist, and its cogency will be admitted by those at least who know the increased power they derive from engaging in continuous and systematic muscular exercise. The gain is the more remarkable in the case of those who exercise their intercostal and abdominal muscles, to the great benefit of their respiration and digestion. For an account of the functional actions of the abdominal muscles, we are indebted to Dr. George McClellan (see his work on Regional Anatomy). That authority points out, in the first place, that the crossed arrangement at the sides of the fibres of these muscles serves to strengthen the abdominal wall, and, when all the muscles of the region act together, they compress and support the viscera and protect them from internal injury. He goes on to say, that "the muscles of the abdomen are quiescent and relaxed during inspiration, but they aid in expiration when the spine is fixed, by drawing the lower ribs downward and inward. When the pelvis is fixed, the thorax is inclined forward by the muscles of both sides acting together : if the muscles of one side act, the trunk is bent to that side. The oblique muscles cause rotation of the trunk, the external oblique turning it to the same side. This is seen in mowing, where the right external oblique and the left internal oblique are simultaneously brought into action. In climbing, the thorax serving as the base of attachment, the abdominal muscles draw the pelvis upward and forward. The chief action of the recti muscles is concerned in flexing the trunk when the pelvis is fixed. Their peculiar segmentation and enclosure in so firm a sheath enable them to maintain their action in all possible bendings of the body."

The muscles of the back are very numerous and strengthen it by five successive layers. Those of the outer layer, the trapezius and the latissmus dorsi muscles, we have already dealt with. The region of the second layer is that of the shoulder-blade and the back part and side of the neck. The muscles of this layer assist in giving movement to the bones of the region. The third layer, which are chiefly respiratory muscles, connect the back ribs with the spine, a longitudinal section traversing the entire length of the back part of the thoracic region. The fourth and fifth layers, which also run vertically, from the dorsal and lumbar regions to the neck, are important muscles in keeping the spine erect, in rotating it, and in giving fixedness to the head and neck. The chief muscle of the fourth layer is the Erector spinæ, which, with its accessories, the longissimus dorsi and spinalis dorsi, serves, as its name implies, to maintain the spine in the erect posture ; it also serves, observes Dr. Gray, " to bend the trunk backwards when it is required to counterbalance the influence of any weight borne in front of the body, as, for instance, when a heavy weight is suspended from the neck, or when there is any great abdominal development, as in pregnancy or dropsy." The other muscles of the back have various and manifold functions. Some muscles, besides giving support to the spine, or acting successively on different parts of it, rotate it ; others again rotate the vertebræ on which the neck and head are poised ; yet others draw the head backwards, or turn it from side to side. Still others, by their costal attachments, depress the ribs, and thus assist in forced expiration. Added to all these functions, is the no less prime one, of giving strength to the back, and, as extensors, of straightening it when bent.

A word, in conclusion, with reference to the chief muscles of the lower extremity—the hip, thigh, and leg. Before passing to these, reference should be made to the muscles situate in the region of the lumbar vertebræ and the pelvis. The chief of these are the PSOAS MAGNUS and ILIACUS, which, acting from above, flex the thigh upon the pelvis, and at the same time rotate the thigh-bone outwards, from the obliquity of their insertion into the inner and back part of that bone. Acting from below, the thigh-bone being fixed, the muscles of both sides bend the lumbar portion of the spine and pelvis forward. They also serve to maintain the erect position, by supporting the spine and pelvis upon the thigh-bone, and assist in raising the trunk when the body is in the recumbent position. The Psoas Magnus muscle extends from the lumbar vertebræ to the upper and inner part of the thigh-bone ; the Iliacus from the inner surface of the pelvic bone to the femur, or bone of the upper leg, near the Psoas. Other assisting agents, in the act of moving the lower limbs and raising the body, are found in the muscles of the gluteal or buttocks region, and on which we sit. The chief of these is the coarse, extensor muscle, the GLUTEUS MAXIMUS, upon which the body specially depends for its maintenance in the erect posture. It extends from the pelvis to the outer part of the thigh-bone. The action of the GLUTEUS MEDIUS from the ilium (the large, flattened pelvis bone) is to abduct the thigh ; while, acting from its insertion in the thigh-bone, it extends the pelvis outwards, thereby assisting in balancing the body when standing on one leg. Its anterior fibres rotate the thigh inwards, the posterior fibres rotate it outwards. The gluteus maximus extends the thigh-bone upon the pelvis, and thus aids the body to rise from the sitting to the erect position. It also aids in propelling the body in running and leaping.

On the inner side of the thigh, connecting the thigh-bone with the front or middle of the pelvis, are a number of important muscles. These are the GRACILIS, PECTINEUS, and the three ADDUCTORS—longus, brevis, and magnus. The gracilis assists the SARTORIUS, the tailor-muscle of the thigh, in flexing the leg and drawing it inwards ; it is also an adductor of the thigh. The pectineus and the three adductors are the chief agents in adducting or drawing close the thigh, as we see in equestrian exercise, the flanks of the horse being grasped between the knees by the action of these muscles.*
These adductor muscles, assisted by the Psoas and Iliacus, flex the thigh upon the pelvis ; in walking, they also assist in drawing forward the hinder limb.

Down the front and outer face of the thigh, run the great muscles, the SARTORIUS, RECTUS FEMORIS, VASTUS INTERNUS and CRUREUS, and VASTUS EXTERNUS. The four latter are usually spoken of as the QUADRICEPS EXTENSOR, or great

---

* For the development of these muscles, Mr. Sandow has invented and patented a machine, to which the reader or young athlete is elsewhere referred.

extensor muscle of the leg. The action of the sartorius is to flex the leg upon the thigh, and, continuing, to flex the thigh upon the pelvis, at the same time drawing the limb inwards, so as to cross one leg over the other. Its position may be traced by the hand, as it passes obliquely across the front of the thigh to the inner side, and then descends vertically as far as the knee, behind which it has its insertion. It is the longest muscle in the body, extending from the pelvis to the inner surface of the tibia, and has the power also of flexing the pelvis upon the thigh, and, if one alone acts, it can rotate the pelvis. The quadriceps extensor extends the leg upon the thigh and straightens the knees. It has a four-headed origin (hence its name) in the anterior, inner and outer surfaces of the femur, near the hip-joint; while its lower insertion is the knee-cap and shin-bone, just below the knee-joint. "Taking its fixed point from the leg, as in standing," says Dr. Gray, "this muscle will act upon the thigh-bone, supporting it perpendicularly upon the head of the tibia (shin-bone), and thus maintaining the entire weight of the body." The rectus femoris muscle, which extends from the pelvis to the knee-cap, "assists the psoas and iliacus in supporting the pelvis and trunk upon the thigh-bone, or in bending it forward."

The muscles of the back of the thigh are the BICEPS, SEMI-TENDINOSUS, and SEMI-MEMBRANOSUS. These are familiarly called the "Hamstring muscles," and their function is to flex the leg upon the thigh. They are peculiar, observes Dr. McClellan, "in that they are too short to allow of full flexion of the hip while the leg is extended. They possess what is called the 'ligamentous function,' owing to their attachment passing over the two joints of the hip and the knee. Thus, when the pelvis is fixed, the thigh can be only moderately flexed while the knee is straight, but as soon as the knee is flexed the hamstring muscles are relaxed and the thigh can be entirely flexed. Acting from below, these muscles serve to support the pelvis upon the head of the thigh-bone, preventing the trunk from falling forward. This is well shown in feats of strength, where the body is thrown backward. When the knee is semi-flexed the biceps rotates the leg slightly outward, owing to its oblique direction downward and outward; and in the same way the semi-tendinosus and semi-membranosus assist the popliteus (the 'ham' or back part of the knee-joint) in rotating the leg inward." The hamstring muscles extend from that part of the pelvic bone on which we rest while sitting to the bones of the leg, the biceps being attached to the fibula, or outer bone, the other muscles to the shin-bone, or tibia.

We now come to the muscles of the front, outer face, and back of the leg proper, that is, from the ankle to the knee. Those in the first group are the TIBIALUS ANTICUS, the thick, fleshy muscle on the outer side of the shin-bone and parallel with it, whose function it is, besides aiding in balancing the body at the ankle, to flex the latter and evert, or turn out, the foot ; the EXTENSOR PROPRIUS POLLICIS, the EXTENSOR LONGUS DIGITORUM, and its tendinous extension, the PERONEUS TERTIUS. The latter and the tibialis anticus are the direct flexors of the instep ; they raise and extend the foot and perform the function of walking. The other muscles act upon the toes, and, with their consorts, aid in holding the bones of the leg in the perpendicular position and give strength to the ankle-joint. The muscles of the outer side of the leg are the PERONEUS LONGUS and PERONEUS BREVIS, which serve to steady the leg upon the foot and aid in maintaining the perpendicular direction of the limb. They also act as the extensors of the foot, thus antagonising the tibialus anticus and peroneus tertius, which are flexors.

The muscles of the back of the leg are found in two layers, those of the surface constituting the strong muscular mass which forms the calf. The latter are called the GASTROCNE-MIUS, the PLANTARIS, and the SOLEUS muscles. Their action is chiefly to raise the body on the toes. With reference to the calf-muscles, Dr. Gray remarks that "they possess considerable power and are constantly called into use in standing, walking, dancing, and leaping; hence the large size they usually present. In walking, these muscles draw powerfully upon the os calcis (the heel-bone) raising the heel, and with it the entire body from the ground; the body being thus supported on the raised foot, the opposite limb can be carried forward. In standing, the Soleus, taking its fixed point from below, steadies the leg upon the foot, and prevents the body from falling forward, to which there is a constant tendency from the superincumbent weight." The deeper-lying muscles of the back leg are the POPLITEUS, the flat muscle that covers the hollow space at the back of the knee-joint, and assists in flexing the leg upon the thigh; the TIBIALIS POSTICUS, the most deeply-seated of the muscles of the leg; the FLEXOR LONGUS POLLICIS, and the FLEXOR LONGUS DIGITORUM,—the former situate alongside the outer and smaller bone of the leg, the latter alongside the shin-bone. These two latter muscles are the flexors of the toes, and, continuing their action, extend the foot upon the leg; they also assist in extending the foot as in the act of walking, or in standing on tiptoe. The tibialis posticus is a direct extensor of the instep upon the leg, and, acting in concert with the tibialis anticus, it turns the sole of the foot inward, antagonising the peroneus longus, which turns it outward. Covering the lower part of these muscles, and extending for about six inches upward from the heel, is the TENDO ACHILLIS, the thickest and strongest of the tendons in the body. The muscles of the ankles and feet need not detain us, our topographical survey of the body, in so far as the muscles benefited by exercise are concerned, having taken us far enough. It is well perhaps to note, before leaving this chapter, that the action of the muscles of which we have been treating may be reversed, according to the part fixed while the individual muscle is contracting.

PLATE VII.

MUSCLES OF THE EXTENDED LEG.

ANTERIOR, POSTERIOR AND LATERAL ASPECTS.

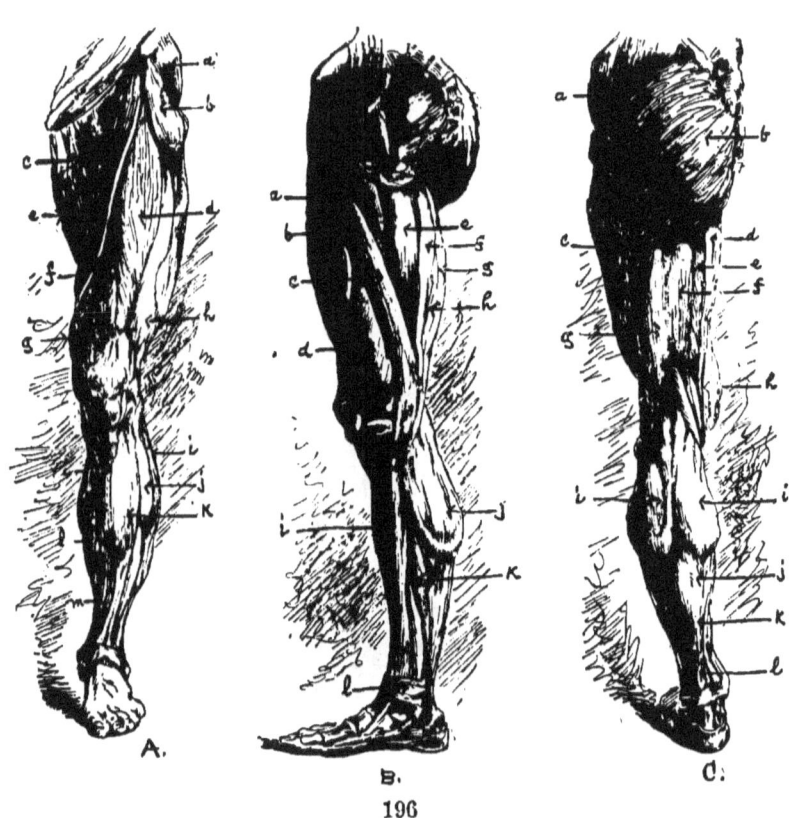

FIGURE, SKELETON, AND MUSCLES OF THE ATHLETE

TO ILLUSTRATE THE DRAWINGS

FROM PROF. C. ROTH'S "ATLAS OF ARTISTIC ANATOMY."

(By permission of Messrs. H. Grevel & Co., London.)

PLATE I.—FIGURE OF THE ATHLETE.

PLATE II.—SKELETON OF THE ATHLETE.

MUSCLES. PLATE III.—ANTERIOR ASPECT.

1. Annular ligament.
2. Flexor longus pollicis.
3. Flexor carpi radialis.
4. Palmaris longus muscle.
5. Pronator teres muscle.
6. Supinator longus muscle.
7. Biceps muscle.
8. Triceps muscle.
9. Coraco-brachialis muscle.
10. Teres major muscle.
11. Deltoid muscle.
12. Pectoralis major muscle.
13. Serratus magnus muscle.
14. Trapezius muscle.
15. Supinator longus muscle.
16. Brachialis anticus muscle.
17. External oblique muscle.

8. Gluteus medius.
19. Gluteus maximus.
20. Tensor vaginæ femoris.
21. Rectus abdominis muscle.
22. Adductor longus.
23. Gracilis muscle.
24. Semi-membranosus muscle.
25. Rectus femoris muscle.
26. Vastus internus muscle.
27. Sartorius muscle.
28. Vastus externus muscle.
29. Gastrocnemius muscle.
30. Tibialis anticus muscle.
31. Soleus muscle.
32. Tendo Achillis.
33. Anterior annular ligament.
34. Fascia lata.

MUSCLES. PLATE IV.—POSTERIOR ASPECT.

1. Extensor carpi ulnaris.
2. Flexor carpi ulnaris.
3. Anconeus muscle.
4. Biceps muscle.
5. Triceps muscle.
6. Tendon of Triceps.
7. Deltoid muscle.
8. Trapezius muscle.
9. Latissimus dorsi.
10. Serratus magnus muscle.
11. External oblique muscle.
12. Gluteus medius.
13. Gluteus maximus.

14. Tensor vaginæ femoris.
15. Rectus femoris muscle.
16. Externus vastus muscle.
17. Gracilis muscle.
18. Semi-membranosus muscle.
19. Internus vastus muscle.
20. Sartorius muscle.
21. Gastrocnemius muscle.
22. Tendo Achillis.
23. Peroneus longus.
24. Tibialis anticus muscle.
25. Tibialis posticus.

MUSCLES OF THE ATHLETE, *Continued*.

A.—ANTERIOR ASPECT OF EXTENDED LEG.

a. Gluteus medius muscle.
b. Tensor vaginæ femoris.
c. Adductor longus.
d. Rectus femoris muscle.
e. Gracilis muscle.
f. Sartorius muscle.
g. Vastus internus.

h. Vastus externus.
i. Gastrocnemius.
j. Peroneus longus muscle.
k. Tibialis anticus.
l. Soleus muscle.
m. Tibialis posticus muscle.

B.—INTERNAL ASPECT.

a. Adductor longus muscle.
b. Rectus femoris muscle.
c. Sartorius muscle.
d. Vastus internus muscle.
e. Gracilis muscle.
f. Adductor magnus muscle.

g. Semi-tendinosus muscle.
h. Semi-membranosus muscle.
i. Tibialis anticus muscle.
j. Gastrocnemius.
k. Soleus muscle.
l. Annular ligament.

C.—POSTERIOR ASPECT.

a. Gluteus medius muscle.
b. Gluteus maximus muscle.
c. Vastus externus.
d. Vastus internus.
e. Semi-membranosus muscle.
f. Semi-tendinosus muscle.

g. Biceps femoris muscle.
h. Gracilis muscle.
i. Gastrocnemius.
j. Soleus muscle.
k. Flexor longus digitorum.
l. Tendo Achillis.

PLATE V.—MUSCLES OF THE TRUNK, SHOULDER, EXTENDED ARMS, AND FLEXED LEG.

PLATE VI.—MUSCLES OF THE FLEXED ARM: ANTERIOR, POSTERIOR, AND LATERAL ASPECTS.

PLATE VII.—MUSCLES OF THE EXTENDED LEG: ANTERIOR AND POSTERIOR ASPECTS.

EXERCISES.

PREFATORY.

BEFORE proceeding to the movements proper, to be detailed in the following exercises, the pupil-in-training would do well to devote some little time at first to a number of free exercises, with the dumb-bells, so as to give suppleness to the limbs, enable the would-be athlete to acquire correct habits of breathing, and accustom himself to easy and well-balanced postures of the body, with due attention to erectness, yet with freedom from rigidity and constraint. The first thing to do is to assume, and practice facility in maintaining the proper standing attitude of the recruit-in-training. This, the commencing position, should be as follows: The heels in line and closed, the knees held well back, and the toes turned out at an angle of 60 degrees. The body full to the front, straight, and inclined forward, so that its weight shall fall on the arch of the instep, supported by the ball of the toes, and only lightly on the heels. The arms should hang tensely from the shoulders, hands firmly grasping the dumb-bells, second joints of the fingers lightly touching the thighs. The hips a little drawn back, the chest advanced, and the shoulders square. The head erect, the chin slightly drawn in, and the eyes looking straight to the front. Regard to this, the proper attitude of the military cadet at "attention," ought to be rigidly enforced in commencing the exercises; for correct habits of bearing the body, when properly acquired, confirm themselves without any exertion, and will add materially to the health and strength of the young athlete.

The great matter to be here attained is, in the case of the young, to quicken the muscular system to a due degree of flexibility, and, in the case of the mature or old, to awaken that which has become stiff or lain dormant, and to train it to become pliant and yielding. We all know the pleasant feeling which we experience when we stretch ourselves when wearied, or when, having sat long in a constrained and unnatural attitude, we have got up and, as we say, shaken or pulled ourselves together, or gone off for a stiff walk. These are Nature's efforts at relaxation, and they can be greatly assisted, and ought to be, by some simple home-exercises, such as those about to be indicated, to relieve and take out the creases from the cramped form. The habit, if constantly practised, of going through these elementary stretching movements, will be found an invaluable one, and the results will be a surprise to many, in the increased suppleness that will ensue and the more perfect command that will be gained over the muscles and the joints. Those who are zealous for the general pursuit of physical culture cannot lay too much stress on these simple and initial exercises, for tney are the first principles in the art of giving mobility and endurance to the human frame. They should, therefore, in all cases, precede the more active exercises, for until you can unstiffen and relax the joints and their connecting muscles and tissues, you can only at the risk of injury proceed with the prescribed physical training. To bring the matter more immediately home to the pupil, let him try at the outset to stoop, without bending the spine, to lace his shoes, touch the floor with his finger-tips, or, keeping his body as erect as he can, bring his toes to his teeth. He will find, if he tries, that a child can beat him at any of these tasks; while, with practice, he will soon be able to rival his infant exemplar; though, of course, he is not expected to become an acrobat or a contortionist. When he has attained this pliancy and increased the contractile power of his muscles, he will have gained much in the functional activity of the body, as well as mastered a pleasurable control over his muscles and joints. Were anything further needed to be said on this topic, it would be this, that without suppleness there is no grace, and the presentable man or woman is not the person whose muscles are atrophied or inelastic, and whose joints are angular or creak.

A little time, as has been said, should be devoted to the free movements, with the dumb-bells, and before entering upon the exercises proper. This will accustom the hands to the grip and weight of the bells. Like putting a rifle into the hand of a soldier at squad-drill, when he has learnt his facings and the goose-step, it will steady the recruit and give resistance and the requisite tension to the muscles, particularly those of the wrist and the forearm. The dumb-bells, it must here be repeated, should, for beginners especially, be of light construction, either of wood or of iron; in the latter case, they may be covered with leather. For women and the youth of both sexes, their weight should range from two to three pounds each; for male adults, from three to five pounds each. The length of time given daily to training must necessarily vary with the age, capacity, and physical condition of the pupil, as well as with the amount of leisure he is at liberty to devote, at any one period of the day, to the movements. If thirty minutes cannot be given continuously to the exercises, perhaps fifteen can be snatched twice a day; but, at the outset, any one exercise should not be prolonged beyond the point when the muscles tire, though every exercise should be continued until they ache, and the mind should be put into the work, that the muscles may feel the strain and receive the full benefit of the toning and building-up process.

This is a point that cannot be too much impressed upon the pupil-in-training, as it is the basal fact upon which all successful physical instruction rests. There must be a concentration of the will-power upon the exercise in hand, and the dumb-bell must be held and used, not passively, but as a potentiality to be actively and strenuously exerted, that the muscles may first be loosened and then alternately contracted and relaxed, in the process which Nature has designed for their healthy growth and development. With flabby muscles there can hardly ever be vigorous frames or sound health. Nor need the possession of either be a matter of serious or difficult attainment. Much might be gained by an exercise of an hour or two a week in the intelligent use of a pair of light dumb-bells. Even out of a daily "constitutional" we might get more benefit did we impart energy to our movements, and put the muscles of progression to strain, in a sharp and exhilarating walk,—bearing in mind that the test of having put the muscles to use is to have tired them.

In giving class-instruction with the dumb-bells, a strict instructor will not allow any lounging about during the lessons. If the lessons are too protracted for the strength of some of the pupils, the latter should be encouraged to continue them as long as possible, but not to overtax their endurance or cause them to lose zest in their work. The exercises should always be returned to with pleasure, and taken up systematically and with eager ardour. Intervals for rest should be frequent, but when they occur, the pupil should be directed "to stand at ease" only, and not to fall out of the ranks, or throw down the dumb-bells heedlessly and without leave. It is hardly necessary to say that no one should be allowed to eat or take refreshments of any kind while the exercises are going on. If the mouth is dry, it may be moistened with a lozenge or confection. Nor should the instructor permit talking among the pupils during the lesson. If directed to perform a movement a certain number of times, they should count under their breath, always breathing freely, but naturally, by the use of the diaphragmatic muscle, which best raises the ribs, expands the chest, and gives freest play to the lungs. Even when putting the muscles to strain during a stiff exercise, the lips should be pressed together as little as possible, the air being inhaled through the nostrils, for the most part, though, in the case of active exercise, respiration may be permitted by the mouth.

In performing the exercises, the pupil, if in the privacy of his own room, will find it less impeding and more comfortable to strip to the waist, or, if in class, to wear a light gymnasium suit, and to spend his strength freely till the muscles tire and the perspiration comes. If possible, let nothing interfere with the time daily devoted to exercise. If this is persisted in, it will soon become a habit, and the pupil will find that if, perchance, he should miss a day's exercise, he will miss it as he misses his bath, and will not feel up to his usual work. The bath, which should be made ready beforehand, should always be taken after exercise, and if the heart is all right and the breathing regular, it may be taken even when heated, though it will be well to let a short interval elapse, so long, meanwhile, as he does not get chilled. The bath should always be cold, the head and breast being first laved with the hand in the water, and then, if it be winter, in for fifteen or twenty seconds and out, or for longer, if it be summer. Keep up a brisk action while in the bath, and when it has been taken, pat rather than rub the body dry.

The preliminary exercises with the dumb-bells may now be entered upon. Those of immediate benefit are the movements tending to give free play to the muscles and joints which, in the later exercises, will be drawn more heavily into service; to relaxing and rendering them supple; and to afford opportunity for acquiring proper methods of breathing under exercise; care being taken to maintain, as far as possible, the erect position and an easy but well-governed control of the body. In breathing, this general rule may be observed, viz.: to inhale the air as the arms are raised or drawn back for action, and to exhale it as they descend or are brought forward to the position of "attention." In squatting or in movements where the body is lowered, the breath should be taken in the downward and expelled in the upward action. In all muscular movements, the action of the lungs in breathing should be kept as free and unimpeded as possible that no strain be felt upon the air passages. All movements should be made evenly and without jerkiness, but with muscles tense and the mind set upon the exercise. Even in the case of the snatching-lifts with heavy weights, the same caution is to be observed, the mind retaining its balance and steady equilibrium as well as the body. In exercising, see that there is an abundance of pure and fresh air, and that the body is unhindered by tight clothing.

The initial exercises with the bells include :—

A. The flexing, or bending, the hand inwards and outwards upon the wrist, and rotating or turning it round, long enough till the muscles ache. These movements will give free play and increased strength to the muscles of the forearm and wrist, add power to the hand, and firmness to the grasp. They should be performed left and right hand alternately, the eyes critically scanning the motions, and the will-power imparting the energy ; then both hands should be exercised simultaneously. Prolonged exercise in this and other movements with the left hand will counteract the tendency to right-handedness and insure a symmetrical development of the body. The fingers can individually be strengthened by lifting the dumb-bells successively with each finger.

B. Keeping the shoulders perfectly square, the body erect, the arms pendant and close to the sides, the hands firmly grasping the dumb-bells, fingers touching the thighs, move the head slowly backwards and forwards, from side to side, then roll it round to the right and left, as far as possible. With eyes to the front, now raise and depress the shoulder-blades and arms, as in shrugging the shoulders ; after which, elevate the arms at full length and in line with the body, and rotate them in both directions until the muscles are tired. These several movements will have a beneficial effect on the respiratory organs and give strength and mobility to the shoulder-joints, as well as to the muscles of the chest and neck.

C. Resuming the attitude of attention, the dumb-bells still in hand, rotate or twist the body on its hip-axis alternately to the left and to the right, keeping the back and the legs straight during the movement ; then sway the trunk on the hips from side to side, bending sideways as far as may be comfortable ; .after which, bend the body backwards and forwards, taking care to keep the legs straight, the chest pressed out, and the head undrooped. These movements will assist the circulation of the blood, as they alternately stretch and shorten the veins, stimulate the organs of the chest and abdomen, strengthen the muscles of the trunk, and give pliancy to the chief hinge of the body, the hip-joint.

D. Toe and heel raising in succession may now be exercised, in which the weight of the body is alternately thrown on the toes and the heels, the body being kept upright, and accommodating itself so far as to maintain the balance. This movement will loosen the ankle-joints, give strength to the muscles of the calf, and accustom the body to preserve the equilibrium. Keeping the body straight and the head erect, knee-bending and stretching may now be exercised, the movement being extended to the squatting position, in which the body is allowed to drop till the buttocks are in contact with the heels (the latter being raised from the ground, the weight of the body resting wholly on the toes), with an alternate quick recover to the attitude of attention. This latter movement brings into play the quadriceps extensor muscle, which extends the leg upon the thigh ; the former movement giving exercise to the muscles chiefly brought into use in the act of walking and the other motions of progression.

Some of these free movements, the pupil-athlete will find, are taken up more systematically in the exercises proper : they are here suggested as a sort of "preliminary canter" or warming-up, before entering on the more serious training-drill which follows. All of them, of course, can be practised *without* the dumb-bells, and may be so recommended as an initial practice for women and children, or for young men of weak constitution and indifferent health, to be afterwards followed, when the frame has been built up, by a course of the exercises proper *with* the dumb-bells.

Before entering upon a systematic course of physical training, the pupil should, to mark the gain in his development, set down the date at which he commenced to practise, and take his height, weight, and the measurements of his chest (normal, relaxed, and expanded), neck, shoulders, forearm, upper arm, waist, thigh, and calf ; and, at stated intervals afterwards, register the increase he has gained, as the result of exercise, and as an encouragement to progress. The height taken should be that without shoes, and the weight that stripped, or in one's usual exercising attire. Of course, the measurements subsequently taken should be that in the attire, whatever it may be, when first measured, and, as far as possible, they should be taken at the same period of the day and after the same amount of muscular exercise. The measurements of the chest and upper and lower limbs should be skin measurements; the chest girth being that well up under the arms, which should be horizontally extended, the line passing over the nipples. The forearm measurement should be that round the thickest part of the extended arm, hands clenched; that of the upper arm over the ridge of the biceps when the forearm is flexed at the elbow. The thigh and calf measurements should be those round the thickest part, when the heels are raised from the ground and the toes are pressed firmly against it, knees well-braced back.

In the following exercises each number is intended to develop its special muscle, or group of muscles; they should therefore be taken up progressively in the order in which they appear. Those who can handle heavier weights than the five pound dumb-bells are recommended to take the simpler exercises with the latter weights first, until they see a visible improvement in their muscles and have trained them to pass to the heavier weights with ease and safety. *All the simpler exercises should be performed with slightly bent knees, that the muscles of the thigh may share in the benefits to be derived from the movements.*

LIGHT-WEIGHT EXERCISES.

Exercise 1.

TAKE a dumb-bell in each hand, and come to the position of attention, as described in the opening sentences in the introduction to these exercises. Now, bend the knees slightly, and turn the inner side of the arms full to the front. In all exercises with the light-weight dumb-bells, the knees must be bent, that the muscles of the leg may feel the strain of the movements of the upper limbs. Tighten the grip of the hands on the dumb-bells, and make tense the muscles of the arms; then alternately flex or bend each arm at the elbow inwards and upwards, till the dumb-bell is in line with the shoulder, back of the hand to the front, shoulders and elbows well drawn down, and the upper arms close to the sides. In lowering the dumb-bells, straighten the arm to its full length, and repeat the alternate movements till the muscles ache.

This exercise will develop chiefly the flexor biceps muscle, and the triceps extensor muscle, of the upper arm.

EXERCISE 2.

This exercise is the same movement as that in No. 1, except that in the position of attention the backs of the hands and the forearms are to the front, and, when the latter are flexed upwards on the elbows, the knuckles of the hands are close to the shoulders. The alternate motion of bending and extending the arm at the elbow is to be performed rhythmically but vigorously, until the flexor and extensor muscles are made pliant and firm. The action will have a stimulating effect on the respiratory organs and the circulation of the blood.

EXERCISE 3.

Come to the position of attention, knees bent as before, and raise both arms outwards, at full length, in a line with the shoulders. Now, turn the inner side of the forearms upwards, and alternately flex each inwards toward the head, until the dumb-bell is immediately over the shoulder. In practising this movement, maintain the arms rigidly in alignment with the shoulders; in other words, do not let them droop; and, in the straightening-out movement, extend the arms fully, and put the muscles to the strain. The chief muscle that comes into exercise here, besides the biceps and triceps of the arm, is the deltoid, the great muscle that caps the shoulder. The effect of these alternate arm-flexings is perhaps more beneficial than when both arms are flexed at the same time. Its chief advantage is that it gives one arm a momentary alternate rest, and does not overstrain the heart by unduly forcing the circulation of the blood.

EXERCISE 4.

This exercise is the same as the last, the flexing movement of the forearms, however, being concurrent or simultaneous, and not alternate. See photo. No. 4, and the caution to be observed, indicated in the closing sentence of the preceding exercise. In the alternate straightening or opening-out movement, care should be taken to extend the arm fully, so that the extensor muscles may have fair play in counteracting the motion of the pulling-up or flexing muscles.

The exercise may be varied with advantage by curving the back slightly and bending the head downwards; at the same time bringing the flexed forearms inwards, underneath the upper arms and shoulders, and in this attitude ply the dumb-bells outwards from the armpits to the full extension of each arm. The exercise will be found beneficial for the biceps, triceps, and deltoid. It will also stimulate the breathing and quicken the blood-currents, to many perhaps the chief need as well as the great advantage of active muscular exercise. In the regular alternation of movements, such as are here and elsewhere in the series indicated, the young pupil should try to observe cadence, for a rhythmic movement tends to the automatic performance of the exercises, and so lessens the sense of fatigue, by relieving the brain of care in directing the muscle-action. The habit, however, of thorough work must be first formed, and the mind fixed on this, before allowing the movements to become automatic.

EXERCISE 5.

From the attitude of attention, simultaneously raise both arms forwards and full to the front, curving them upwards until the hands and dumb-bells meet together in a line with the mouth, elbows straight, head well back. The dumb-bells in this exercise should be held perpendicularly, not horizontally. From the position attained, simultaneously throw

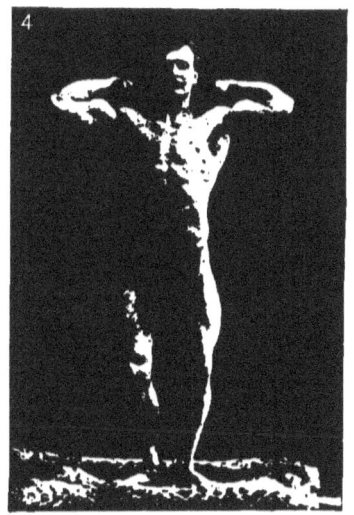

Sarony—Photo.

SANDOW. LIGHT-WEIGHT DUMB-BELL EXERCISES: FIGS. 1 TO 4.

both arms smartly back, well to the rear, and in a line with the shoulders, chest well out. Return them quickly to the front again, and repeat the opening-out movement as often and as vigorously as you can. This exercise is designed to open out the chest, and to loosen and give mobility to the pectoral muscles of the chest, and those in the region of the shoulders. It will be found to have a blood-relieving effect on the organs of the chest and head. Two photographs, Nos. 5 *a* and *b*, illustrate the exercise.

Exercise 6.

From the position of attention, flex both forearms upwards from the elbow, palms inwards, as shown in the left arm of photograph No. 6. Now, alternately raise each arm in a vertical line with the body, taking care to extend the arm over the head to its full length. The return movement should bring the elbow back close to the side and well to the rear. The head and trunk should be kept straight, the chest pressed forwards, and the arms kept well back, during this movement. The muscles brought into play in this exercise, in addition to the biceps, triceps, and deltoid, are those of the back and sides, chiefly the trapesius, latissimus dorsi, and teres major. Their action tends to open the chest and increase its mobility.

Exercise 7.

Take the position of attention; the hands and dumb-bells resting lightly on the front of the thighs, knuckles outwards, knees bent, chest drawn inwards and downwards, back slightly curved. Raise the arms alternately, stretched to their full extent, forwards and upwards, till they are in a line with the top of the head, lowering the one arm as the other is raised. Maintain this exercise as long and as briskly as possible, taking care to leave the lungs and breathing action absolutely free and unimpeded. The movement tends to increase the mobility of the shoulder-joints, and especially to strengthen the anterior deltoid, the serratus magnus, latissimus dorsi and pectoral muscles.

EXERCISE 8.

This exercise will be found useful for loosening and making flexible the muscles of the wrist. From the position of attention, elevate both arms outwards until they are at right angles with the body, keeping them rigid and the muscles tense. Then, turn each hand and dumb-bell simultaneously round as far as possible on the axis of the wrist, maintaining the movement till pliancy is imparted and the muscles are tired. The exercise may be supplemented by bending the hand backwards and forwards on the wrist. See that the arms do not droop from the shoulder alignment, and that they are not allowed to turn on the elbow-joint: the movement is wholly executed by the wrists.

EXERCISE 9.

Take up the dumb-bells by the sphere or bulb ends, grasping the bosses firmly in the hollow of the hands. Now, simultaneously raise the arms outwards, in a line with the body, till they reach the level of the shoulders. In this position, rotate the right-hand dumb-bell from left to right, and the left-hand dumb-bell from right to left, by a circular motion of the wrist. Keep up the exercise till the muscles tire. The rotary movement is executed wholly by the wrist, and will be found to act beneficially on the numerous muscles of the forearm, and tend to give them mobility.

Morrison—Photo.

SANDOW. LIGHT-WEIGHT DUMB-BELL EXERCISES; CHEST-EXPANDING EXERCISE.

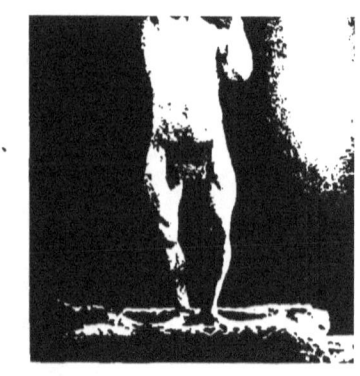

Sarony—Photo.

SANDOW. LIGHT-WEIGHT DUMB-BELL EXERCISES: FIGS. 5B, 6, 7 AND 8.

EXERCISE 10.

This is the same movement reversed ; that is to say, the rotary movement of the right hand should be from right to left, and of the left hand from left to right. The continued exercise of this movement will give flexibility to the muscles of the forearm, and impart to them strength and endurance.

EXERCISE 11.

Place the dumb-bells on the floor, where they should lie lengthwise along the outer side of each foot, the centre of the bar on a line with the toes. Seize them and rise to the position of attention, the head and body erect, the knees unbent. Turn half round on the heels to the left, the toes being at an angle of 60 degrees ; the body, which should turn on the hips, ought, as much as possible, to keep the front position. At the same time, bring the left forearm upwards to the waist, at right angles with the body ; take a good step forward with the right foot, and lunge out forcibly with the left arm in the same direction, as if striking a hard blow, and recover quickly. Bring back the advanced leg with the alternate recover. Repeat the movement until the muscles are well exercised, the right arm remaining tense by the side. In the return movement, bring the elbow well back, and press the chest well forward. The muscles brought into play in this exercise are the anterior deltoid, the biceps, the triceps, the serratus magnus, and the pectoralis major. When the body is turned on the hip, the lunging movement is beneficial to the abdominal muscles and assists circulation in that region.

EXERCISE 12.

This is the same movement reversed ; bringing into play

the right arm and left foot instead of the left arm and right foot. The half-turn will consequently be to the right, and the left foot be advanced, to maintain the balance of the body. In these movements circulation and respiration are materially benefited. The breath should be inhaled as the arms are drawn back, and exhaled when thrust forward. This and the previous exercise, it will be noted, vary from those which precede them, in this respect, that the pulling-up muscles have hitherto been exercised, while the stretching-out ones come now into play.

EXERCISE 13.

This exercise is practiced without the dumb-bells. From the position of attention, the pupil will throw himself forward towards the floor, supporting the body, in a rigid position, on the unbent arms and the toes; then, alternately lower the body, by slowly bending the elbows, until it reaches the prone position, and raise it, by straightening the arms, repeating the movement as many times as possible. Care should be taken that the body and lower limbs are kept rigidly straight and do not touch the floor, that the head is kept well up and the knees unbent. The exercise will be found excellent for strengthening all the muscles of the body, and for expanding the chest. As the strain, in the dipping and raising of the body, is severe, the movement should be indulged in mildly, until the biceps and triceps are pretty well hardened. When facility in the movement has been gained, the effort should be made to stretch the body, in the prone position, horizontally forwards as far as possible (nose more in front), at each performance, that the full benefit of the exercise may be obtained.

EXERCISE 14.

This exercise is the same as No. 13, only rendered more

Sarony—Photo.

SANDOW. LIGHT-WEIGHT DUMB-BELL EXERCISES: FIGS. 9, 11A, 11B AND 12.

difficult by the tension of the rubber straps which encircle the neck, and, by the resisting power, increase the development of the arms in the effort to raise the body from the prone position. The exercise will be more fully explained, with a description of the machine, to which the rubber straps are adjusted, in a later page. See front and profile views in photographs Nos. 14 *a* and *b*.

EXERCISE 15.

This exercise is designed to bring into play the rectus abdominis and other muscles of the abdomen, and has an important effect on the digestion. It should at first be performed without the dumb-bells, then with dumb-bells of increasing weight. Lie flat on the back on the floor, couch or bench, covered by a rug, at full length, the arms close by the sides, the feet pushed under the bureau, weighted or strapped to the floor, to give purchase to the body; then alternately raise the body on the hip-joints, from the prone to the sitting position, and slowly lower it again to the horizontal position, repeating the movements until the abdominal muscles feel the tiring effect of the exercise. After a time, when the pupil has accustomed himself to the strain of the movement, he may render it more difficult by taking a dumb-bell in each hand, and, when in the prone position, raising the arms and stretching them back over the head, and then going through with the trunk-raising and lowering movements, as above described. The exercise may also be performed without weighting or strapping the feet. A deep breath should be taken before raising the body, and exhaled in lowering it. In raising the body to the sitting position, it should also be bent forwards as far as possible towards the feet. With the dumb-bells in the hands, it will be found advantageous also to cross the wrists over the head, and so bring the body upwards and forwards, the head locked, as it were, in the upward-extended arms and moved in unison with them. For persons of full habit and having a tendency to be fat, the exercise will be found very beneficial, the increased blood-circulation absorbing the fatty deposits, and the exercise itself being unfavourable to fatty formation.

EXERCISE 16.

This is a squatting exercise designed to develop the quadriceps extensor, or great extensor muscle of the thigh. (See page 192.) Take a dumb-bell in each hand, and come to the position of attention, the body straight, the head erect, the chest thrust out, and the shoulders and hips held well back. By bending the knees, dip the body in a vertical line to the heels, keeping the back straight and the chin drawn in. Recover and repeat the movement until the muscles ache. This is a good exercise in poising the body and in giving suppleness to the knee-joints. If the muscles of the leg and thigh have been well toned, their natural elasticity will render the movement easy. Take care not to let the body sway or incline forwards or backwards on the hips. After a pause, the exercise may be varied by raising the heels and throwing the weight of the body entirely on the toes, keeping rigidly the position of attention, and rising as high as possible in each motion without losing the balance. Continue the movement for some time, as it will be found of much benefit to the muscles of the calf; it will also give elasticity to those of the foot and ankle.

EXERCISE 17.

This exercise may be practised either with or without dumb-bells. From the position of attention, slowly bend the trunk outwards on the hip-joint, alternately to the left and right,

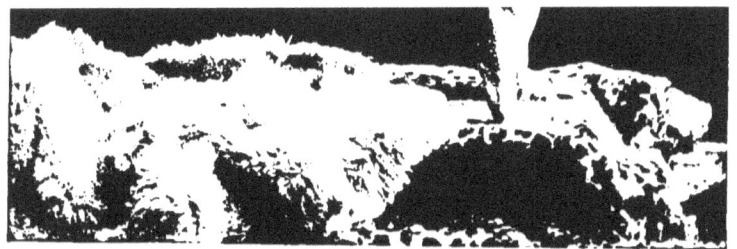

Sarony—Photo.

SANDOW. CHEST-EXPANDING EXERCISES: FIGS. 13A AND 13B.

SANDOW. CHEST-EXPANDING EXERCISES WITH MACHINE: FIGS. 14A AND 14B.

the hand or dumb-bell slightly pressing the outer side of the thigh, and slipping down until it reaches the bend of the knee. When one hand touches the side of the knee the other hand should be raised under the arm just above the serratus magnus muscle. The exercise will be good for the balancing muscles of the trunk as well as for the obliquus abdominis and other muscles that support and protect the sides of the abdomen. It will also give flexibility to the back-bone, and increase the blood circulation, chiefly along the feeding veins of the stomach and the liver.

NOTE.—It has been thought well to append here, *by way of suggestion*, the following table giving the number of times the movements in each of the foregoing exercises are to be practised daily, and the ratio of increase on each occasion afterwards, as the pupil may feel himself able to bear the strain of the more heavily-imposed task. Women and children should try to do one-fifth, or one-fourth, the number of movements indicated for men.

Ex. No.				
1.	50 times, each hand.	Increase every day,	5.
2.	25 " " "	" " "	2.
3.	10 " " "	" " "	1.
4.	10 " " "	Increase every third day,	1.
5.	5 " " "	" " other "	1.
6.	15 " " "	" " " "	1.
7.	10 " " "	" " " "	1.
8.	till arm drops		
9.	" " "		
10.	" " "		
11.	10 times.		
12.	10 "		
13.	3 "		
14.	2 "		
15.	3 "	Increase every second day	1.
16.	10 "	" " " "	1.
17.	15 "	" " " "	1.

HEAVY-WEIGHT EXERCISES.

INTRODUCTION.

THE exercises in Heavy-Weight Lifts, it must here be said, chiefly by way of caution, are designed for those only who desire, and have the necessary strength, to become athletes. For ordinary health purposes, and for reducing corpulency or checking the tendency to become fat, the light-weight exercises which precede those now about to be detailed, will be found sufficient, especially in the case of those who have not robust frames, or whose daily life limits them to confining pursuits. Heavy-weight lifts, of course, should not be attempted by those who suffer from spinal complaint or have weak hearts, though both ailments are hygienically benefited by a course of exercise with the light-weight dumb-bells. To those who feel strong enough for the task, however, and who, by the loosening and hardening of the muscles gained in the previous

Morrison—Photo.

SANDOW. EXERCISES FOR THE MUSCLES OF THE ABDOMEN: FIGS. 15A, 15B AND 15C.

exercises, have acquired facility in handling weights, the following movements may be indulged in, though with dumb-bells ranging, it is suggested, at first, from 12 to 56 lbs., and, afterwards, beyond those, to weights always within the strength-compass and adroitness of the athlete. He will soon learn, not only what weights are within his ability safely to lift, but how to balance the body in the line of gravity, that the weight may be poised with the support of the whole frame, rather than with the muscles of the arm alone. It is recommended that the pupil, before proceeding to the heavy weights, should spend at least three months in performing the preliminary light-weight exercises.

EXERCISE ILLUSTRATED BY PHOTOGRAPHS 18. AND 19.

HOW TO LIFT BY ONE HAND FROM THE GROUND TO THE SHOULDER.

Place the dumb-bell longitudinally between the feet, sphere-ends to the front and rear, the connecting bar of the bell—which should be 4¼ inches in length—in line with the hollow of the foot, the heels ten inches apart, and the toes turned out at a comfortable angle. (See photo. No. 18.) In lowering the body to grasp the dumb-bell, bend the knees, but keep the back straight. Grasp the dumb-bell with the right hand, the arm straight, the left hand resting on the forepart of the left thigh. Without pausing, pull the dumb-bell straight up to the chest, using the left thigh as a fulcrum; at the same time, flex the forearm at the elbow, and straighten the knees. The instant this is done, dip the knees smartly, and, by a simultaneous motion, turn the bell upwards by getting the right forearm underneath it, the elbow resting on the hip-joint, the left hand at ease on the left hip. (See photo. No. 19.) This exercise will be found beneficial to the biceps of the arm, and to the lower limbs, the latter contributing two-thirds of the requisite energy. The movement may also be performed in the same manner with the left hand, the right hand giving the purchase on the right thigh.

To elevate the dumb-bell from the shoulder over the head, the movement may be performed either by the jerk or by the slow-press motion : the latter mode will be described in the next exercise. To elevate by the jerk, dip the knees smartly, and throw the arm upwards to its full extension, bringing the bell over the head, in the centre of the body's gravity. In these one-hand exercises, especially, the eyes should follow the movements of the hand-encircled dumb-bell. The two movements described in this exercise may be made continuous, though performed in two time-beats ; *one*, from the ground to the shoulder, *two*, from thence to the full extension of the arm over the head. The muscles benefited by raising the bell aloft are the chief muscles of the whole body—those of the shoulder, arm, back, chest, and legs.

EXERCISE ILLUSTRATED BY PHOTOGRAPHS 20 TO 24.

ONE-HANDED SLOW-PRESS FROM THE SHOULDER.

The pupil-athlete will observe that the photos. Nos. 20, 21, 22, 23, and 24 form one group, illustrating the slow-press movement successively from the shoulder to the full extension of the arm over the head, photo. No. 22 being the rear view of the attitude illustrated by photo. No. 21. In the successive movements, the eyes, as we have previously said, should follow closely the hand-encircled dumb-bell, that the body may poise itself in concert with the slow raising of the right arm, and so maintain the proper equilibrium. The weight of the bell must depend upon the skill and capacity of the pupil to raise it; he should try to raise as much as he

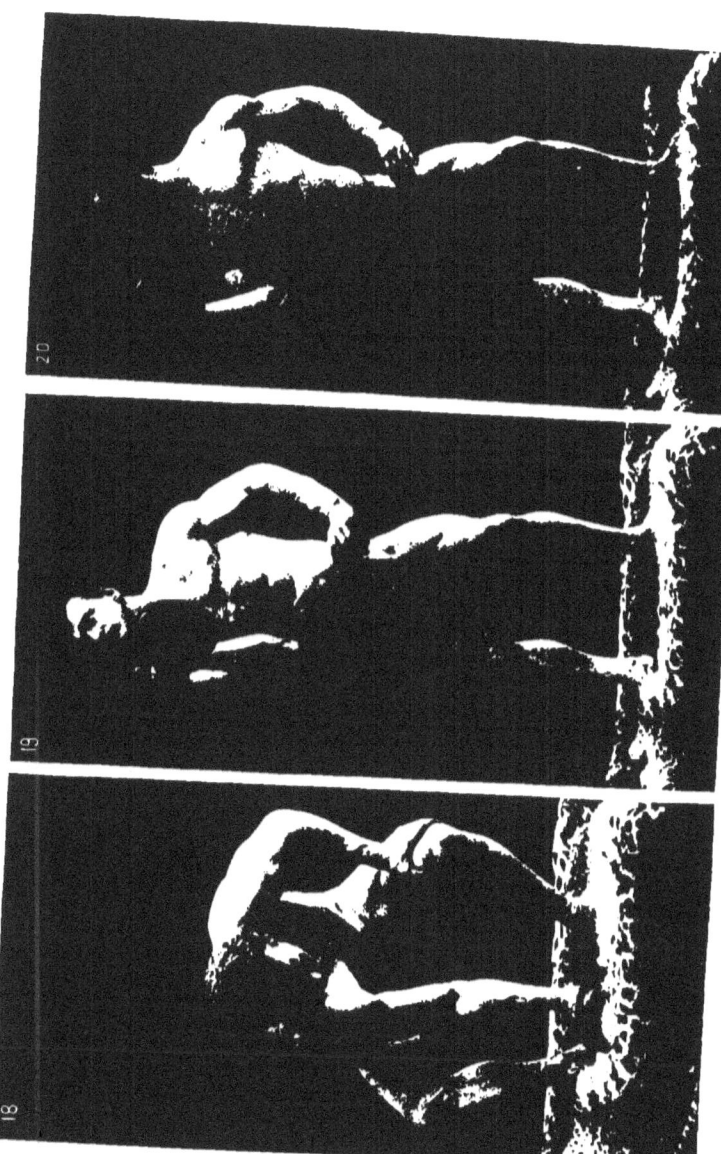

SANDOW. HEAVY-WEIGHT EXERCISES. HOW TO LIFT BY ONE HAND FROM THE GROUND: FIGS. 18 AND 19. ONE HAND SLOW-PRESS FROM THE SHOULDER: FIG. 20.

comfortably can, and increase the weight slightly as he gains in strength and dexterity. The dumb-bell is lifted from the ground to the shoulder as in the previous exercise, the forearm when flexed being held a little more out from the body. To raise the weight from the shoulder by the right hand, the body should be inclined over to the left, the left arm pressing against the upper-left thigh. As the arm is pressed upwards, the body should curl downwards and to the right, until it gets underneath the weight, the upper arm receiving partial support from the latissimus dorsi muscle and arm-pit. By a strenuous effort, continue the up-pressing motion, which will be materially assisted as the body is straightened, aided by the pressure of the left hand upon the left thigh. The action of the disengaged arm and hand should be carefully noted by the pupil-athlete. To make the matter clear, it may be observed, that much assistance in the slow-pressing aloft of heavy weights is rendered by the arm and hand not actively engaged in raising the weight. When the dumb-bell has been raised half-way up, in the righth and, the left forearm, which has been resting on the upper thigh, must now be instantly replaced by the left hand, the latter continuing the pressure on the thigh, helpful in straightening the body and aiding the right arm to elevate the weight. When curling the body under the dumb-bell, keep the forearm which presses it aloft always perpendicular by deflecting it outwards, so as to maintain the true vertical position. The feet, as a rule, should never change place in these slow-press exercises from the shoulder.

The above slow press exercise from the shoulder may be performed with the left hand, though, of necessity, with lighter weights, to prevent injury to the heart, which, in all these left-handed movements, should not be put to an undue strain. This is Nature's own caution, though we may not violate her laws by encouraging ambidexterity, and utilising, more than we do, the left hand. In these, and suchlike exercises, the pupil will find that he can press aloft a heavier weight than he can jerk up, and can, correspondingly, raise more by the jerk from the shoulder than by the snatch-lift from the ground aloft to the full extension of the arm. The gross weight raised by the jerk, is governed, in a measure, by the weight of the body, which must act as a counterpoise to the weight of the bell, otherwise the body will fall over; while the gross weight raised by the snatch-lift is, in part, governed by the power of the hand to grasp the weight. Sandow's highest record in snatch-lifting, from the ground over the head, is 186 lbs.; his weight-record in jerking upwards from the shoulder, is 212 lbs.; in slow-pressing aloft from the shoulder, his record is, for the left hand, 300 lbs., and for the right hand, 322 lbs. In the above exercise the muscles benefited, besides those of the arm and shoulder, are the muscles of the back, loins, and lower limbs.

EXERCISE ILLUSTRATED BY PHOTOGRAPHS 25 AND 27.

ONE-HAND SWING-LIFT FROM THE GROUND OVER THE HEAD.

The first position in this exercise varies from the usual attitude of attention. It is that shown in photo. No. 25, the pupil standing over the bell, head bent and eyes looking down, the right hand about to grasp the dumb-bell, the left ready to place on the corresponding thigh for support. The dumb-bell, it will be seen, is placed on the floor between the feet and well to the rear—the fore-lying sphere being in line with the heels, which should be further apart than in the previous exercises. The object of placing the dumb-bell behind the feet is that the necessary swing may be given it in the curved forward and upward movement, as the hand seizes it and elevates

Sarony—Photo.

SANDOW. HEAVY-WEIGHT EXERCISES. FIGS. 21, 22, 23 AND 24.

it aloft, the left hand resting meanwhile on the thigh, which acts as a fulcrum. The bar of the dumb-bell, in these swing-lifts, must be grasped close to the fore-lying sphere until the weight is swung well up, when, by a slight jerk upwards, the centre of the bar and the proper poise are gained. The advantage of this is obvious, as the upper sphere of the dumb-bell will be supported in the lifting movement by the grip of the closed thumb and fingers, while the lower sphere, swinging free, will, by its own weight, receive greater impetus in the ascent. The pupil will now put the movement into practice, taking care to keep the back as straight as possible, bending the body freely on the hips, and, as the bell curves upwards, incline the body backwards, and move the right foot a little further to the rear, to preserve the balance. The elevating of the dumb-bell aloft, it will be understood, is a continuous movement, the right arm getting under it when it has been swung up from the floor, by a quick dip of the knees, and the instantaneous straightening of the arm and left leg, the left arm bracing the body by the support given the hand on the left hip. The exercise will be good for strengthening the spine, and the muscles of the chest, arms, and lower limbs.

Exercise 28.

SLOW LIFT FROM THE GROUND TO THE SHOULDER.

This is a slow lift from the ground to the shoulder, designed chiefly to develop the biceps and forearm. Photo. No. 28 will illustrate the first position, the dumb-bell being placed transversely between the feet, the right hand grasping the bar, the inner side of the forearm and the clasped fingers to the front, the left hand braced on the left fore-thigh. From this position, pull the bell steadily up as high as the knees and slowly curl it forwards and inwards to the shoulder, by flexing the forearm on the elbow and bending the wrist inwards as much as possible, the hip-joint acting as a fulcrum. Repeat the movement several times, alternately with the right and left hands, and let the weight drop slowly down to the floor.

Exercise 29.

SWING RING-AND-BALL LIFT FROM THE GROUND OVER HEAD.

This is an effective as well as graceful exercise, calling into play the chief muscles of the trunk and limbs, and imparting litheness and elasticity to the movements. The bell is placed on the floor a little in front of the feet, ring to the right, heels in line, and about ten inches apart. Bending the body on the hips, now stoop and grasp the ring by the right hand, knuckles to the right, then pull the ball up sufficiently to clear the floor and swing it inwards between the legs, then, as it swings outwards again, bear it aloft, taking a step forwards at the same time with the right foot to give purchase to the movement and balance to the body. As the ball gains the elevation of the head in the ring-grasped hand of the outstretched arm, tilt it to the back of the hand, by an adroit turn of the wrist, at the same time thrusting the arm fully out, as in the act of lunging upwards, the body being thrown forwards to assist, by its weight, in pressing the ball up, and especially to ease or break the force of the contact of the ball on the forearm, as it is tilted to the back of the hand in the upward ascent. Repeat the movement, which will be found an exhilarating exercise, observing the caution not to injure or break the forearm by permitting the ball to come rudely into contact with it as it is swung aloft. Photo No. 29 illustrates this exercise.

SANDOW. HEAVY-WEIGHT EXERCISES. ONE HAND SWING-LIFT FROM THE GROUND OVERHEAD. FIGS. 25 AND 27.

Morrison—Photo.

Sandow. Heavy-weight exercises.
Fig. 28. Slow lift from ground to shoulder.
Fig. 29. Snatch ring and ball lift from ground overhead.
Figs. 30 and 31. Two-handed lift from ground to shoulder.

EXERCISE ILLUSTRATED BY PHOTOS NOS. 30 AND 31.

TWO-HANDED LIFT FROM THE GROUND TO THE SHOULDER.

The photos No. 30 and 31 will illustrate the successive attitudes in the performance of this exercise. Place the dumb-bells close to the outer side of each foot, the body, in an erect position, standing over them, the heels closed, the toes turned out at a comfortable angle, the head bent and the eyes directed downwards, the arms pendant, but held out a little from the body ready to grasp the bells. Keeping the back straight, by bending the body on the hip-joint and the legs at the knees, stoop down and grasp the dumb-bells close to the front bosses, as in photo. No. 31. Now, with a quick movement, pull the bells straight up to the sides of the chest, in line with the arm-pits, elbows bent outwards, the movement being aided by a hard pressure with both legs; then step smartly to the rear with the right foot, and, slightly bending both knees, turn the bells upward with a sudden jerk, and get the forearms underneath them, the elbows resting on the hip-joints. The whole movement is a quick one, the legs bearing the chief strain. To elevate the bells from the shoulder, the movement can be practised either with the jerk or with the slow-press motion. The jerk movement is much the same as in elevating by one hand: practically it is easier, as the two weights maintain the body in equipoise. To elevate by the slow-press, the weight of the body must be thrown on the rear leg, which may be drawn further back to give increased purchase, as the dumb-bells rise, and to preserve the balance of the body. When half-way up, slow-press the weights firmly and bring the upper part of the body under the dumb-bells: this will make the weights easier to press, and be a good strengthening exercise for the spine.

EXERCISE ILLUSTRATED BY PHOTOS. NOS. 33 AND 34.

HOLDING OUT AT ARM'S LENGTH WITH BOTH HANDS.

This is a holding-out exercise, to give strength and endurance to the arms and back, and to develop the muscles of the chest and shoulders. The front and rear views of photos. Nos. 33 and 34, illustrate the exercise. The dumb-bells are elevated over the head, to the full extension of the arms, as in the previous exercise. Now let them fall slowly down and outwards, till the upper arm is in alignment with the shoulders, twisting the forearm partially to the rear and bending the shoulders backwards to give increased support in bearing the weights. Let the dumb-bells be as heavy as the pupil can safely use, increasing the weight as strength and facility are gained. The exercise can be varied by bringing the bells from the elevated position slowly down and out in front of the body, knuckles upwards, and in a line with the mouth. Maintain the position as long as possible and replace the bells on the floor. The latter exercise will be good for the deltoid, trapesius, and latissimus dorsi muscles—that is, for the shoulder-muscles and those of the upper chest and back.

Sarony—Photo.

SANDOW. HEAVY-WEIGHT EXERCISES: FIGS. 33 AND 34, HOLDING AT ARMS' LENGTH; FRONT AND BACK VIEWS.

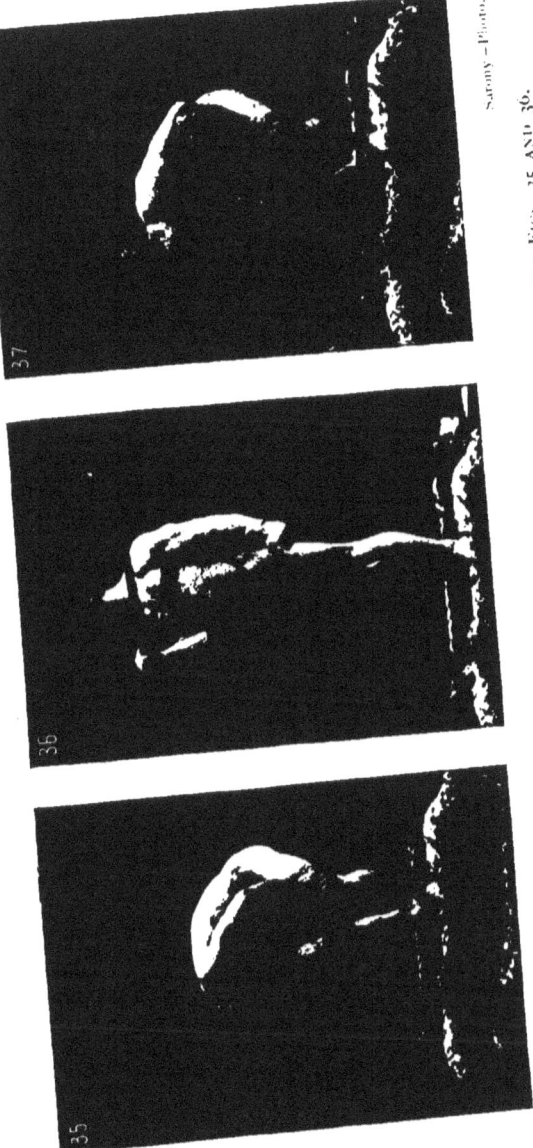

SANDOW. BAR-BELL EXERCISES. ONE HAND LIFT FROM GROUND TO SHOULDER: FIGS. 35 AND 36.
SNATCH LIFT FROM GROUND OVERHEAD: FIG. 37.

BAR-BELL EXERCISES.

ONE-HANDED LIFT FROM THE GROUND TO THE SHOULDER.

Illustrated by Photos. Nos. 35 and 36.

IN these one-handed lifts the centre of the bar should, by some device, be indicated, to mark readily the place to be grasped, so that a perfect balance may be obtained. This is the more important, as no time should be lost in the tiring stooping attitude preparatory to grasping and elevating the bell. In stooping, keep the back as straight as possible, by bending the body on the hip-joints and the legs at the knees. The bar-bell should be placed horizontally on the ground, the centre of the bar over the instep. the heels together, and the toes turned slightly outwards. The right hand will now grasp the bar-bell, the inner side of the forearm to the front, and as straight as possible, the left hand resting on the left fore-thigh, near the knee, thumb inside and fingers outside the leg. Now pull the bar straight up as high as the waist, the upper arm close to the body, the forearm at right angles with it, momentarily resting on the hip. At this instant, take a step to the right rear with the right foot, and, by bending the knees, turn the bar upwards by a swift movement of the wrist, getting the forearm underneath it, then press up to the shoulder, recovering the right foot and straightening the body. From the shoulder, the bar-bell may be elevated aloft, either by the jerk or by the slow-press movement, as in the methods described in raising the heavy-weight dumb-bells. While at the shoulder, the bar-bell, however, should be turned round at right angles to the body, spheres to the front and rear, and steadied, the eyes following the movements of the hand, that the proper balance may be maintained and the body suffer no strain. The same movement may be gone through with the left hand and a lighter weight, thus developing both sides of the body symmetrically. The muscles benefited in this movement are those of the shoulder, chest, and legs, as well as the pulling and stretching muscles of the arm.

EXERCISE 37.

ONE-HANDED BAR-BELL SNATCHING LIFT FROM THE GROUND OVER-HEAD.

The first position in this exercise is that described in the previous one, with this difference, that the backs of the forearm and hand, in grasping the bar, are to the front. It is a one-handed snatch-lift from the ground to the full extension of the right arm over the head. The whole exercise should be performed in one movement, without pause, the backward step being taken to maintain the balance, as the body recovers

SANDOW. BAR-BELL EXERCISES. EXERCISE FOR BOTH HANDS: FIGS. 38A AND 38B.

Morrison—Photo.

the upright position. The first motion, which merges at once into the second, should bring the bar, by a rapid snatch up along the body as high as the shoulder, when, by a sudden dip of the knees, the right arm should get underneath the bell, and, with a quick pressure of the legs, give it the needed impetus to the first motion to speed it aloft. This is an excellent exercise for the legs, right arm, and back : its practice with the left hand is also recommended, so as to develop both legs and arms equally. If you let a weight down slowly with one arm to the ground, hold the other straight out from the body to preserve the balance.

BAR-BELL EXERCISE FOR BOTH HANDS.

ILLUSTRATED BY PHOTOS. NOS. 38 *a*, *b*, *c*, AND *d*.

Bar-bell exercises should be performed with progressively increasing weights, according to the strength and dexterity of the pupil. The two-handed movement will bring into play all the muscles of the body and upper and lower limbs, especially those of the forearm and wrist, and will be found beneficial in expanding the chest and promoting circulation and digestion. Photos. Nos. 38 *a* and *b* will show the correct position of the bar-bell on the ground and the first attitudes to be assumed by the pupil. The bar-bell is placed squarely in front, across the instep of each foot ; the body straight, the arms held a little out in front ; the hands ready to make the grip. Now, stoop from the waist, or bend the knees, keeping the back straight, and seize the bar with both hands, knuckles to the front, the hands being from 16 to 18 inches apart, according to the height and breadth of the pupil. With a swift motion, raise the bar upwards and outwards, letting it turn in the hands, as the forearms are flexed at the elbow and placed under it by a quick dip of the knees, and bring it in a line with the shoulders, palms to the front, as in photo. No. 38*b*., the knees being straightened by a simultaneous movement, and the left foot carried six inches to the rear to preserve the balance. To raise the bar-bell over the head to the full extension of the arms, the movement may be done with a jerk, the knees, by a sudden dipping motion, giving spring to the movement. Hold the bell aloft for a moment or two, as a test of endurance, or, if of a comparatively light weight, repeat the elevating movement. When exercising with weights with arms stretched above the head, always let the weights come down slowly, that the triceps muscle of the arm may feel the developing strain of the movement. With a bar-bell of heavier weight, the elevating movement from the shoulder over the head should be done by the slow-press motion, the legs as well as the arms participating in the movement, and contributing their share of support. By the same motion the bell may be gradually lowered to the chest, and then replaced on the floor.

Exercise 39 and 39*a*.

SLOW BAR-BELL LIFT FOR DEVELOPING THE MUSCLES OF THE FOREARM AND WRIST.

This is a slow-lift exercise, designed to benefit chiefly the muscles of the wrist and forearm. Photos. Nos. 39 and 39*a*, show the mode of turning the bar in the hand, by a slow movement, as it is brought from the thigh to the waist. Practice in this turning movement, which should at first be performed with a light-weight bar-bell, will strengthen the wrist, and enable the pupil to acquire the knack of the twist, preparatory to pressing the bell up to the top of the chest. From the attitude of attention, bend the body on the hip-joints, keeping the back as straight as possible, the arms close to the side, and the heels together. Now grasp the bar-bell with both hands,

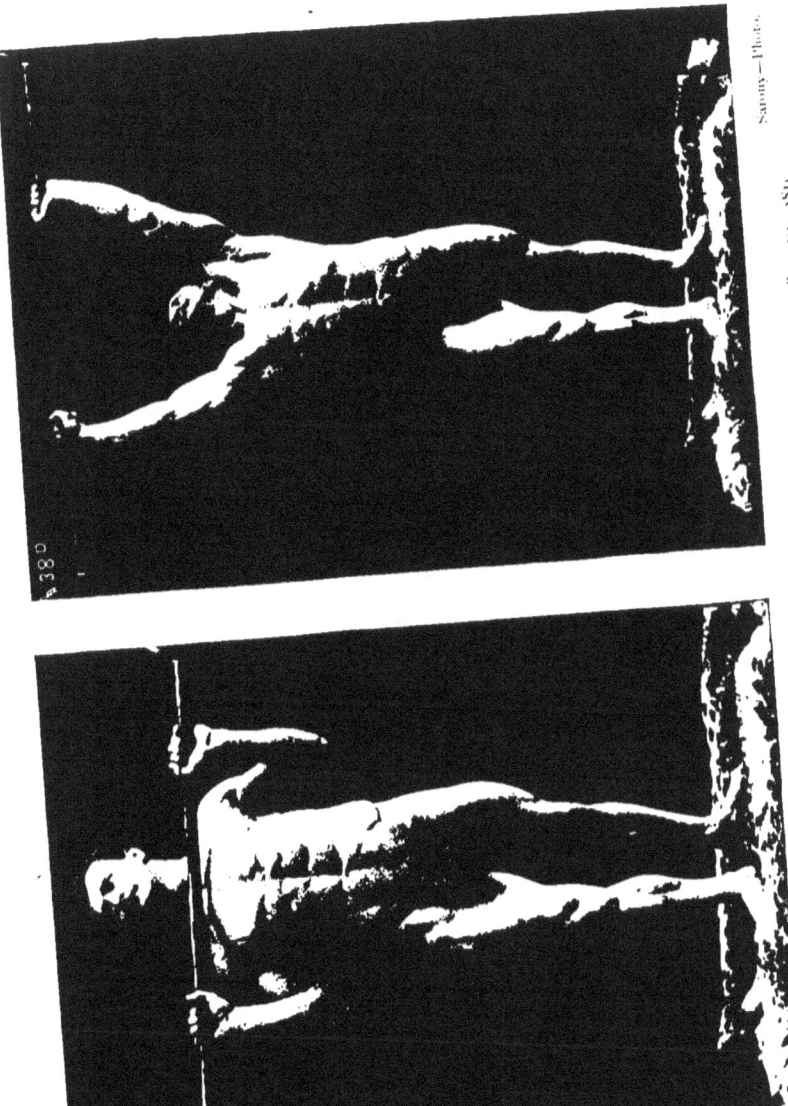

SANDOW. BAR-BELL EXERCISES. EXERCISE FOR BOTH HANDS: FIGS. 38C AND 38D.

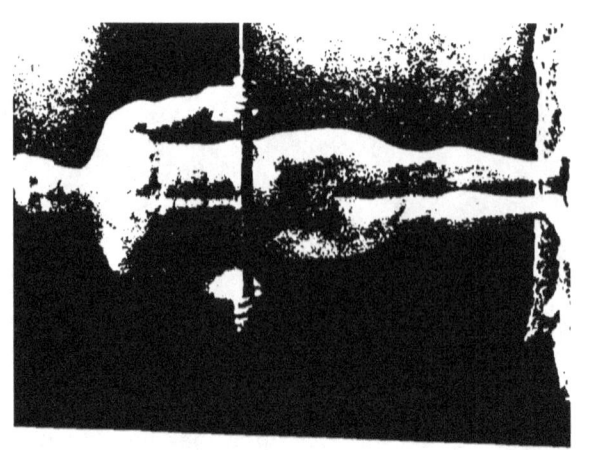

knuckles to the front, and pull it steadily and slowly up to the thigh, and straighten the body. The position is illustrated in photo. No. 39. From the thigh, raise the bar-bell slowly outwards and upwards, by bending the forearms at the elbows, and the hands backwards on the wrists. The bar in this position will be clasped by the hands, the weight resting chiefly on the thumb, and the first joints of the turned-in fingers, as in photo. No. 39 a. Lower the bar to the thigh, and repeat the movement, as a practice to the wrists. To elevate it from the waist to the top of the chest, continue the pressure of the forearms from the elbows, until they are well underneath the bar, then press slowly up. The exercise may be continued by elevating the bar-bell from the chest, above the head, to the full extension of the arms, or over it to the rear, to be afterwards lowered to rest on the nape of the neck and the shoulders. When raising the bell aloft from the chest, do not bend the back; stand perfectly straight and keep the head erect This is a good exercise to repeat, as it will give flexibility to the shoulder-joints, and develop the chest and the pushing muscles of the arms.

EXERCISE 40 AND 40a.

ONE-HANDED BAR-BELL LIFT, UPRIGHT POSITION.

This exercise is another mode of bringing the bar-bell to the shoulder, and may be practised as follows: The bar-bell, instead of being placed horizontally on the ground, is placed on end, resting on one of the spheres. It may be raised either by one hand, or by both, to the shoulder, according to its weight and the ability of the pupil to wield it. Photos. Nos. 40 and 40 a. illustrate the two initial positions. To raise the bar-bell with one hand, grasp it firmly with the right hand in the centre of the bar, bending the body and the knees as little as may be necessary. Now push the lower sphere outwards, and, as the upper sphere tilts over, balance the bar on the upturned palm and raise the bell to the shoulder by the pressure of the forearm, making a lever with the elbow on the hip, the pressure upwards being aided by the straightening motion of the body and the knees. From the shoulder, the bar-bell may be raised overhead by the jerk or by the slow-press motion, taking care, in either act, to keep the eyes on it so as to maintain the poise of the bell and the balance of the body.

Exercise 41.

TWO-HANDED BAR-BELL LIFT TO THE SHOULDER, UPRIGHT POSITION.

To raise the bar-bell with both hands from the upright position on the ground to the shoulder, stoop down and grasp it firmly with both hands, as in photo. No. 40a., tilting the upper sphere over the shoulder to the rear, the body and feet adapting themselves to the swaying and steadying motions; then by a firm pressure push it up to the shoulder. When this position has been gained, aided by the left hand in raising the weight to the shoulder, the bar-bell will rest entirely in the right hand, grasped round the centre of the bar, and the left hand will be withdrawn. When the bell is properly poised, it may be elevated, as before, to the full extension of the uplifted arm, by the jerk, or by the slow-press movement. If the bell be of moderate weight, the exercise may with advantage be repeated, as it will be of benefit to all the muscles of the body, as well as to those of the upper and lower limbs.

Exercise 42.

FINGER-LIFT FROM THE GROUND.

This is an exercise which the pupil must adapt for himself, using any article which may fit itself to the purpose and can

SANDOW. BAR-BELL EXERCISES. UPRIGHT POSITION: FIGS. 40 AND 40A.

Morrison—Photo.

be caught up on the crooked finger, such as a chair, a portmanteau, or a scuttle of coal. The weight, which practice will enable the pupil successively to increase, may be suspended on the inner joint of the middle, or other, finger, at arm's length from the body, or raised between the legs, the young athlete having first placed his feet on two strong and firm chairs, or any platform raised above the elevation of the article to be lifted. Mr. Sandow's record-weight for finger-lifting is 600 pounds. In raising this weight, he usually stands on an elevated staging, over a frame and platform, upon which rest the men or material designed to be raised. In all heavyweight lifting, care should be taken to keep the back straight, to prevent strain or rupture, and to throw the chief pressure on the legs. In right hand lifts, the left hand should find purchase by pressing against the left thigh, and *vice versa*.

EXERCISE 43.

ONE AND TWO HAND STONE-LIFTS FROM THE GROUND.

A stone lift from the ground may be raised in the same manner as described in the previous exercise for finger-lifts. Photos. Nos. 43 and 43 *a* will illustrate the position, the athlete standing astride the weight to be raised, his feet planted on fixed benches or steady chairs, on either side of the weight. It will be found convenient to use straps round the wrists that will not slip over the hands, but aid the latter in the grasp and pull of the weight. The weight should be raised by a straight pull upwards, the back being kept perfectly unbent, and the body not too far lowered to miss the purchase which the legs afford in the uplifting and straightening movement. When the weight is raised by one hand the disengaged hand will gain support by resting on the complementtary thigh. It will be usually found that the athlete can raise more by one hand than he can raise by two, the disengaged hand lending material assistance in the weight-lifting process. Mr. Sandow's stone-lifting record is 1,500 pounds.

EXERCISE 44.

HARNESS-AND-CHAIN LIFT FROM THE GROUND.

Photograph No. 44 will illustrate the position assumed in heavy-weight lifting in harness. A strong, broad collar, it will be seen, is placed round the neck and over the shoulders, to which are attached four suspended chains, with hooks at the ends, to be fastened to the weights in the stooping attitude preparatory to raising them. When the collar has been adjusted, and the proper position taken up, stoop down with a straight back and fasten the hooks, then place both hands on the thighs, and by a firm pressure of the legs force the body upwards. The exercise will be good for the shoulders and back, and especially for the straightening muscles of the legs and arms. Mr. Sandow's record for harness-lifting is 4,800 pounds.

HARNESS AND CHAIN LIFT: FIG. 44.

SANDOW'S PHYSICAL TRAINING LEG MACHINE.

In the previous pages we have more than once referred to this ingeniously contrived and useful machine, designed and patented by the great athlete, with the object of providing the necessary apparatus for exercising the lower limbs. With the bar-bells, and the dumb-bells, of heavy and light weight, the leg machine is the only mechanical appliance which Mr. Sandow uses or finds essential to his simple and efficient methods of physical training. It completes and rounds off his system of muscular exercise by bringing into play (1), the extensor and flexor, that is, the stretching and pulling-up muscles of the leg, and (2), the abductor and adductor muscles, viz., those muscles that separate or draw apart, and bring together again, the lower limbs. The adductor muscles of the leg, more popularly speaking, are those which we use in gripping the sides of a horse in equestrian exercise. It is these abductor and adductor muscles that Mr. Sandow, with his accustomed thoroughness in seeking to develop the *whole* body, and not parts of it merely, has had in view to exercise by means of this invention, for these muscles of the inner and outer thigh, which supplement and re-enforce those used in the act and motions of progression, usually come little into play. The value of the machine will be better appreciated if one reflects on the fact that the customary movements of the legs, if one is not a horseman, are chiefly forwards and backwards, as in walking, running, jumping, rowing, and bicycle-riding; while the lateral movements are little, if at all, exercised, and the muscles situate on the inner and outer thigh are neglected or kept dormant.

The leg-machine, which is of simple design and comparatively cheap in its construction, is so made as to be easily taken apart, packed up, and, when desired, transported from place to place. The illustrations, Nos. 45 *a*, *b*, *c*, and *d*, will show its design and uses, while a previous illustration (Nos. 14 and 14 *a*), referred to in Exercise No. 14, exhibits another adaptation of the invention in developing the muscles of the arms, shoulders and back. The machine consists of a base-board or platform, from five to six feet in length, having at either end an upright post or standard, secured by screws to the baseboard, and capped by ferrules with attached hooks or eyes, and a cross-bar for the hands to rest upon and give steadiness to the upright posts. About the middle of the cross-bar or brace, and a little apart, are two fixed hooks upon which are hung stirrups, connected by one or more rubber straps or elastic cables; into these stirrups the feet are placed for the purpose of exercise, either by a direct up-and-down tread, or by alternate lateral thrusts to the outer base of the machine.

To the hooks on the top of the upright posts are fastened single, double or treble cables, which are attached at the other end to strong leather straps, padded on the inside. These straps are buckled round the legs, below or above the knee, so as to

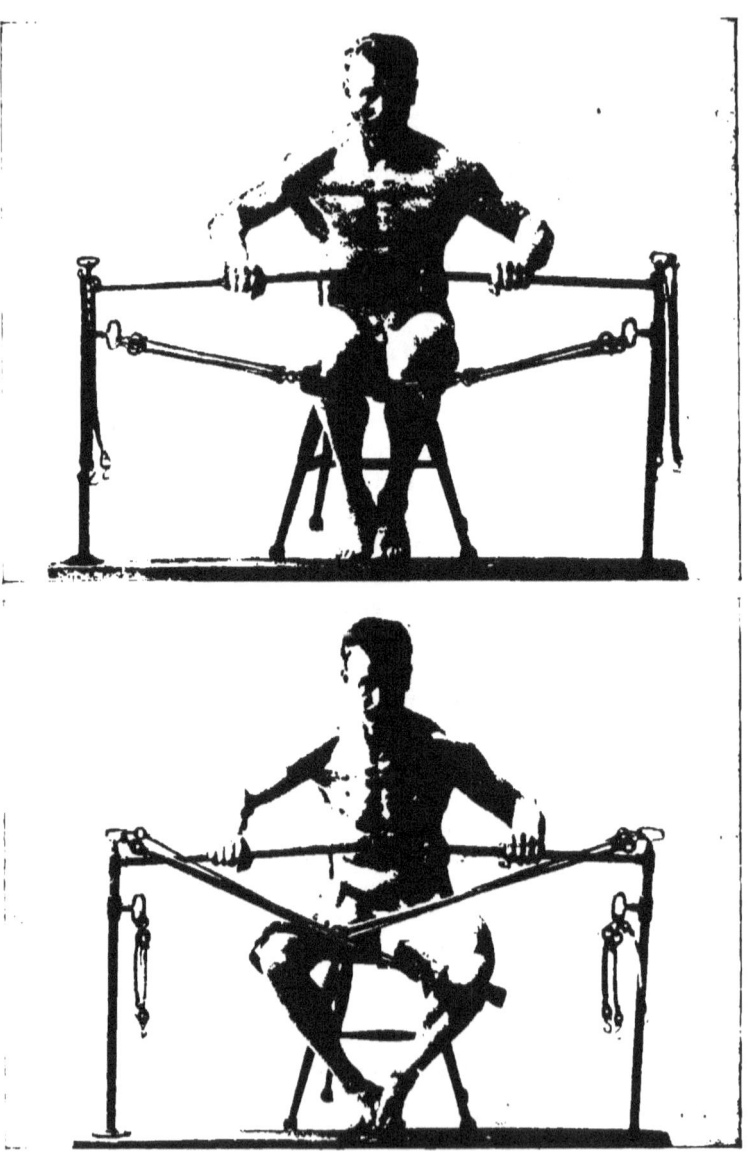

Merrison Photo.

SANDOW. LEG MACHINE EXERCISES: FIGS. 45A AND 45B.

Morrison—Photo.

SANDOW. LEG MACHINE EXERCISES: FIGS. 45C, 45D AND 45E.

exercise the abductor and adductor muscles. The cables pull the separated legs together, as shown in illustration No. 45b, and the exercise is derived by stretching the legs apart and allowing the cables to pull them slowly together again. The position of the pupil in this exercise is that shown in the photo., seated on a chair, hands clasping the brace, heels together, toes alone resting on the platform and aiding the limbs to press themselves apart. The movement should be repeated as long as the operator cares to give to the exercise; it will be found good for the sartorius and the triceps muscles of the leg. If one cable coupled to each leg is not sufficient of a strain, then two or more may be used. In this exercise, the cables should cross each other and hook in the straps of the far leg, one being fastened above and the other fastened below the knee.

A little distance below the upper end of the standards are additional hooks, to which are attached shorter elastic cables, provided at the further end with snap-hooks, to be attached to the outer side of the padded straps that encircle the legs just below the knee, (see illustration of the operator, No. 45a). In this exercise the position of the operator is much the same as in that of the previous exercise, with this difference, that the knees are brought together by a strong pressure and allowed slowly to be pulled apart by the tension of the rubber cables, the movement being good for developing the biceps muscles of the leg.

APPENDIX A.

TABLE PREPARED BY PROF. W. O. ATWATER, TO ILLUSTRATE THE AMOUNT OF PROTEIN AND ENERGY OBTAINED FOR 25 CENTS IN FOOD MATERIALS AT DIFFERENT PRICES PER POUND.

The following figures, which are based on analyses and prices of specimens of materials purchased in New England and in New York City, will illustrate the variations in the amount of nutritive material obtained at the same cost in different food materials at different prices.

	Protein. Grams.	Energy. Calories.
Beef, sirloin, 25 cts.	68	870
Beef, sirloin, 20 cts.	86	1114
Beef, neck, 8 cts.	218	2795
Mutton, leg, 22 cts.	77	1075
Salt pork (bacon), 12 cts.	9	7295
Chicken, 22 cts.	127	695
Salmon, 30 cts.	54	520
Salt cod, 7 cts.	259	1105
Oysters (40 cts. per quart), 20 cts.	36	325
Hen's eggs (25 cts. per dozen), 18½ cts.	77	910
Milk, 7 cts. per quart. 3½ cts.	109	2180
Cheese, whole milk, 15 cts.	213	3420
Butter, 30 cts.	none	3080
Sugar, 5 cts.	none	9095
Wheat flour, 3 cts.	418	13680
Wheat bread, 7½ cts.	136	4255
Corn (maize) meal, 2 cts.	518	20230
Oatmeal, 5 cts.	345	9190
Potatoes, 75 cts. per bushel, 1¼ cts.	163	7690
Standards for day's food for laboring man at moderate work. Voit's (German)	118	3050
Standards for day's food for laboring man at moderate work. Writer's (American)	125	3540

—From *The Forum*, Sept., 1893.

DIRECTIONS FOR READING THE SANDOW ANTHROPOMETRIC CHART.

A brief explanation of the chart may be given as follows: The *horizontal* lines extending across the chart represent the parts of the body measured, the names of which are given at the sides. The *vertical* lines give the percental values of the different measurements, ranging from the minimum at 0 on the left to the maximum, 100, on the right.

The figures at the top show, by percentages, the relative values of the heavy vertical lines, and the intervening light lines divide these spaces into four equal parts, making each subdivision *between* 10 and 90 per cent, 2¼ per cent in value, but *outside* these points only 1¼ per cent.

The figures *above* indicate the per cent of individuals who were found to surpass and the figures *below* the per cent of those who failed to surpass any given point.

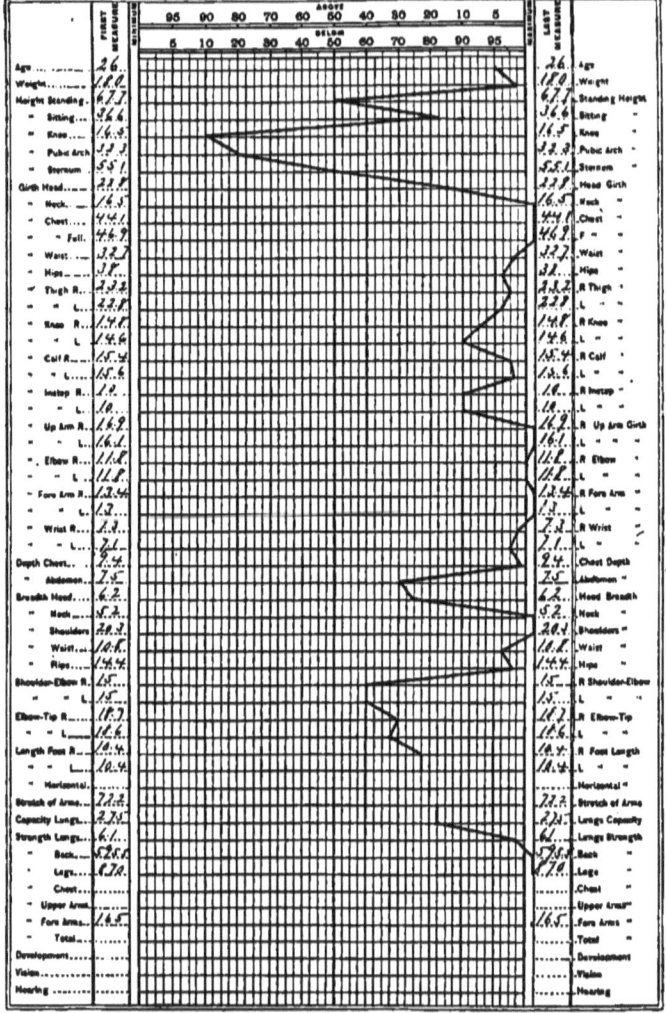

DR. SARGENT'S CHART SHOWING MR. SANDOW'S MEASUREMENTS AND THE VARIATIONS FROM THE NORMAL (THE VERTICAL DARK LINE IN THE MIDDLE OF THE CHART). SEE NEXT PAGE.

APPENDIX C.

TABLE OF THE INCREASE IN THE MEASUREMENTS

of a pupil of Mr. Sandow (Mr. Martinus Sieveking, of Chicago), within the space of three months' practice with heavy-weight dumb-bells, on the great athlete's system of Physical Training. (See photograph of pupil). The result has been achieved, it is proper to state, after Mr. Sieveking had gone through the preliminary course of light-weight exercises, with six-pound dumb-bells.

Weight, 175 pounds (increase, 15 pounds); height, 5 feet 11 inches. Measurements: Neck, 18 inches; chest, 43 inches (increase, 3 inches), chest expansion, $7\frac{1}{2}$ inches (increase, $3\frac{1}{2}$ inches); biceps, $16\frac{1}{4}$ inches (increase, 2 inches); forearm, 15 inches (increase, $1\frac{1}{2}$ inches); waist, 26 inches (reduction, 3 inches); thigh, 23 inches (increase, 2 inches); calf, 16 inches, (increase, $1\frac{1}{2}$ inches).

MR. SANDOW'S COMPETITIVE PRIZE AWARDS.

It is Mr. Sandow's design to award a prize in each city or town he visits in which to give his public exhibitions, to the individual who, on furnishing adequate proof, has gained most within a given period under his system of physical training by the use of light and heavy-weight dumb-bells. Personal communication with Mr. Sandow will elicit the precise conditions on which it is intended to give these awards to pupils-in-training. On this subject, and with regard to the agency and sale of Mr. Sandow's patent Physical Training Leg Machine, dumb-bells and bar-bells, communication should be made to Mr. Sandow, care of his manager,

MR. F. ZIEGFELD, JR.,
PULLMAN BUILDINGS, CHICAGO, ILL.

SIEVEKING, A PUPIL OF SANDOW'S.

APPENDIX D.

The striking result of four months' training, according to Sandow's methods, on a delicate Eton boy: a letter from Captain Greatorex, Assistant-Inspector of Military Gymnasia for the British Army.

The Gymnasium,
Aldershot, 9th January, 1894.

Dear Mr. Sandow,—

You may perhaps consider the following case worthy of insertion in the book you are shortly publishing, as an instance of the results accruing in a very short space of time to an individual by the persistent following out of your system of light dumb-bell exercises, etc.

In July last I was asked if I could suggest any means of improving the physique of an Eton boy, who was under the required chest measurement for the army, *i. e.*, for admission as a cadet to the Royal Military College, Sandhurst.

Being an old pupil of yours, and having great faith in your system (when the pupil has a real desire to work and improve his physique), I determined to see what it would do in this instance.

I subjoin my young friend's measurements taken by me on the 25th July, and again on the 26th November. The results are wonderful, and speak for themselves. Yet this is not a fair test of your system, for I was only able to give him *ten* lessons.

When he first commenced, he could not press off the floor *once*, but after the expiration of four months I saw him execute this exercise 37 consecutive times, and he *now* does it 150 times *each* day.

In July last he was, to use a slang term, a terrible "weed," but now is a fine, smart, upstanding young man—with pronouncedly good and erect carriage of body—and a general air of pride in his own manhood. The coats he now wears will not button across his chest by many inches.

He wrote me from Cologne a week ago. His weight is now 10 stone 7 pounds, a gain in five months of $17\frac{1}{2}$ pounds. I will take fresh measurements when he returns to this country, and send them on to you.

Instead of being *much below* the average physique, as he was in July last, he is now *much above* it, and rapidly developing into a very fine young man. I wish you to distinctly understand that for these four months he has had *no* time to devote to other physical exercises, recreative or otherwise, than yours, as he has been working very hard for the Army Entrance Examination. The average time he has been able to give to his exercises has been *half* an hour *twice* daily.

You will, I am sure, agree with me that this young gentleman deserves very great praise for the dogged aud persistent way in which he has worked; for, however good the system, it is null and void without the concentrated "will-power" of the pupil upon the work in hand.

With best wishes for the New Year, and hoping soon to see you back in England,

Believe me,

Faithfully yours,

F. W. GREATOREX, Capt.,

Assistant-Inspector of Gymnasia.

To Professor EUGENE SANDOW,

New York, U. S. A.

Date.	Age.	Weight.	Horizontal Measurement of Chest, full.	Horizontal Measurement of Chest, empty.	Right Biceps.	Right Forearm.	Right Deltoid.	Left Biceps.	Left Forearm.	Left Deltoid.	Remarks.
1893.	Yrs.	Stones	In.	In.	In.	In.	In.	In.	In.	In.	
July 25.	18½	9.3¼	35½	32	11	9⅞	14	10¼	9⅝	13¾	These measurements were taken after 3 weeks' continuous work.
Nov. 26.	18 10/12	10	38½	34¾	12	10½	15	11⅝	10	14¾	
Increase......		10¾ lbs.	3	2¾	1	⅝	1	1⅜	⅜	1	

www.ingramcontent.com/pod-product-compliance
Lightning Source LLC
Chambersburg PA
CBHW031330230426
43670CB00006B/297